The Engineer's Diet

LUNCH
SPECIAL!
ENGINEER'S
PLATE

The Engineer's Diet

Thomas J. Shaw

Illustrated by Don Berry

Foreword by Jimmie Shiu, M.D.

Manto Press

Berkeley, California

First Printing, 1998

Manto Press
1442-A Walnut Street, Suite 390
Berkeley, CA 94709

Manufactured in the United States of America

∞
Printed on archival quality, acid-free paper

ISBN 0-9664856-1-0
Library of Congress Catalog Card Number 98-66463

Disclaimer:
The ideas, procedures, and suggestions contained in this book are not intended as a substitute for consulting with your physician on medical matters. Individuals with medical conditions should consult physicians regarding any significant lifestyle changes.

Contents

Foreword

MY RESPONSE WAS FAVORABLE to *The Engineer's Diet* almost from the first page. Not only did I recognize at once the basic common sense of what Thomas was saying, I was also entertained. This is one of those books that makes you laugh out loud while you're reading.

Thomas Shaw has been at the center of innovation in health care for more than a decade. I got to know him a few years ago in connection with his breakthrough invention of an automated retraction syringe to prevent needlestick accidents. I can say without reservation that he is a man with a mission. If a thing can be done, he won't quit until he figures out how to do it.

The Engineer's Diet is not only worth reading, it's worth trying. For the many who try and try, and fail, to lose weight, Thomas provides a refreshing and completely innocuous approach. From a medical standpoint, there are no harmful side effects. The way I see it, what have you got to lose—besides pounds? I'll go further and recommend this book to people who *don't* have a weight problem. It's important to all of us to read good ideas, and this book is full of them.

As a physician, I agree with Thomas that we should eat naturally, letting the body tell us what it needs. I have never thought it was smart to restrict the body to certain foods, particularly when it comes to high- or all-protein diets, which may be risky. As he analyzes and compares leading diets of the day, Thomas really does us a favor, pointing out that some of these highly specialized eating plans are contradictory to normal body function. Those of us trained as doctors are used to the idea that the body is set up to balance itself. From organic chemistry, we know about the Krebs cycle—the process by which the body converts carbohydrates, proteins, and fats to produce energy. Your body actually is able to neutralize or convert whatever it needs to regulate both excesses and deficiencies. In short, Thomas is right about common sense already built into us by Nature.

I applaud his thinking also on the role that metabolic rate plays in weight loss and health in general. Popular mythology notwithstanding, there are differences between what kids can do, what 20- to 30-year-olds can do, and what older adults can't do even if they want to. When Tom says that as adults we don't need all that extra food for energy, he's absolutely on target. For one thing, what and how much we eat determines the body's nitrogen content, nitrogen being a main element in building tissue. When we're young and developing, the nitrogen balance is positive. As we get older, we require less nitrogen. To put it simply, when this shift occurs, our tissue-building capabilities slow down, and eventually they stop. Then, no matter what we take in, we can't convert energy to tissue—for example to build up new muscle. Naturally, the timing of this change differs among individuals, but all of us will reach this stage.

I am not an expert on exercise. But anyone knows that too much exercise will overtax the system. Remember back when folks first started all this jogging?

Everybody thought it was the greatest. But I was worried that people might forget how our bodies are actually put together. What about our knees and ankles, and other joints? We really aren't built to run like some other animal species, and Thomas is right, overexercising has definite risks. He makes an interesting point also about conserving the heart by "firing up" the digestive system less often. That is, by skipping lunch!

Thomas's attention to our emotional and mental health is timely and insightful, given the preponderance of health care professionals who recommend or prescribe drugs for anxiety and depression. I agree that medication in this country is overdone. How did we manage to roam so far from the natural world? Can we find our way back to common sense? Thomas reminds us that people didn't always live or eat the way Americans do today. It's a certainty, for example, that manual laborers and farmers who rose early and ate in the morning, and then worked hard all day before eating supper, did not suffer in great numbers from obesity.

All Thomas Shaw asks is that we stop to think, to investigate, to question more deeply. I could talk about many other points that Thomas makes in this book, but it's better if you read it for yourself. Given the engineer's perspective, his ideas become clear and logical, relevant indeed to the modern world's most overweight society, and funnier the further you read.

JIMMIE SHIU, M.D.
Fellow of the American College of Surgeons
Dallas, Texas, 1998

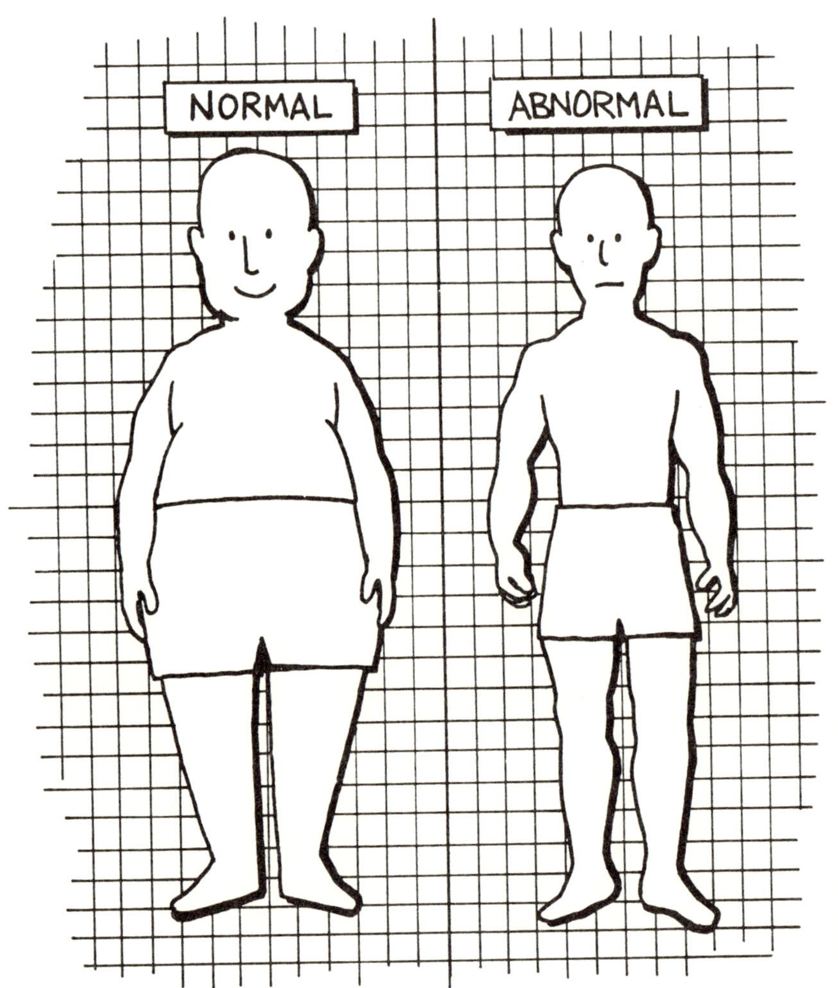
NORMAL
ABNORMAL

Preface

LET ME START BY CLARIFYING a few things. This book is not written for people who don't have a weight problem. It is not a fitness book. It is not a nutritional guide, or a cookbook either. It won't provide you with therapy or tips on how to dress to match your real self, or overcome your shyness. It won't teach you new ways to cope. I'm tired of self-help books. I'm tired of theories of self-control. I'm tired of the "science" of weight management. And I'm damn tired of tasteless food. I wrote this book for people who, like me, are overweight just from eating a normal, 20th-century American diet.

I recognize that there are individuals living in this country who, because they are young, or because of some other special circumstances, are not overweight. Well, this book is not for them. A person who lives in this country, eats at least three times a day, and who is *not* overweight, is definitely not the norm. You've heard the saying "If you're normal in an abnormal situation, then you're not normal." In the current U.S. scene with our overly plentiful supply of high-calorie food, our largely sedentary lifestyle, and our custom of eating three meals a day, if you're not fat, then for sure you're not normal.

Don't be confused by what you see around you. If you're watching television and reading health and fashion magazines for guidance on how you should look and feel, you're making a big mistake. Most people who have come up with the "latest and the best" eating plans and fitness programs are not normal and they're certainly not telling the unvarnished truth about themselves. They're not admitting to what, under other circumstances, might be classified as a disorder—regarding food or exercise or both.

I'll say just a few words to those of you in the "skinny" category who happen to be reading this book. Watching what you eat, controlling your portions, or denying yourself what your body wants are basically unnatural behaviors and could be regarded as early warning signs of a potential eating disorder. When you find yourself exercising steadily enough to "raise your metabolism so it will burn better" long term, some sort of treatment is indicated, and I suggest you seek help. If you are over 25 and not overweight and not a professional athlete, your body, in contrast to what you believe, is relatively inefficient.

Think about it. As a constant consumer, you are actually totally dependent on modern society even though you may be highly compulsive about controlling your intake. The fact is that, for whatever unfortunate reason, you're not capable of storing energy. You would, for example, be one of the first to die in the event of a famine. In times of plenty, you have either a psychological problem that keeps you from being able to eat and enjoy food that attracts normal people, or a physical disability that makes it impossible for you to gain weight, even when food is right in front of you. You may or may not derive benefit from reading this book, but be kind enough at least to spare us your opinions on how to manage our weight.

Now, to address the rest of my readers. You belong to a majority class of healthy people with average willpower and no particular mental problems. But you are overweight precisely because you eat the normal American way. You are the people for whom I wrote this book. I sincerely hope that as we look around us, realizing just how normal we are and that those others are caught in an unfortunate plight from which escape will be difficult, we will be kinder to skinny people than they have been to us.

Finally, let's clear up a common confusion about what losing weight really means. Up to now, you may have believed that it's hard to reach your desired weight. That is simply not true. The fact is, you already have reached and surpassed it. You've achieved your goal with pounds to spare! All you have to do is get rid of the surplus. Ask yourself this question: Supposing you wanted to make a small fortune. Would you rather start with no money and have to build up to it? Or start with a big fortune and cut it down? You tell me which is easier.

Acknowledgments

I want to thank Dr. Jimmie Shiu for suggesting that I write down my experience to help others. Warmest thanks to my editor and book doctor, Rebecca Salome, who told me if the first draft hadn't made her laugh out loud, she wouldn't have stayed up nights to work on it. Special thanks to my mother, Elizabeth Shaw, for superb copyediting. Three cheers to all of the people who have gone on The Engineer's Diet at great personal loss (of weight, that is). I would also like to thank

Aimee, of Aimee's Coffee Haus in The Colony, Texas, for providing me with my new, favorite noontime hangout. Last but not least, I would like to thank my wife, Sue, and my children, Steven and Rebecca, for supporting me emotionally during all my big, fat, grumpy years before I solved my weight problem.

Introduction: My Story

IN ONLY ONE WAY IS THIS LIKE other diet books. It's about losing weight. By cutting calories. Doctors write about it; nutritionists too. And even some fitness experts. Sounds simple enough, but look around. So far, it isn't working too well. Everybody knows what to do, so why do we still have the weight problem? That's where I come in. I'm an engineer by profession. As an engineer, I'm responsible not only for figuring out what needs to be done, but for getting it done the best way. That may be the biggest difference between me and the doctor. Engineers don't just have to build bridges; we have to build them so they don't collapse—ever! We experiment too, but preferably in the lab, not over the water and loaded with traffic!

Eating less is what this book is all about. I think it's important that I'm an engineer because I know a little something about engineering aspects of the human body—things like weight and load of materials, fuel efficiency, optimization, maintenance over time, depreciation, and so on. My purpose in this book is simple. I want to share with others the logical thinking that set me free from 27 years of being overweight. My story is probably not much different from yours, if you're an overweight American like most of us. Over many uncomfortable years, I experienced periods of low self-worth, lack of confidence, an intermittent sense of doom, and constant awareness of the unhealthiness of my lifestyle. At times I briefly overcame these feelings through dedication to work and family, but at no time was I ever unaware of my extra, undesirable self. My blood pressure was elevated, my ability to exercise was severely compromised, and I never lost the constant sense that I had to succeed quickly in business so I could find the time and energy to address my weight problem. I can't help but believe that my experience is shared to some degree by many of my overweight American brothers and sisters. Some of us have resolved to accept the condition as a part of modern life; many have tried to use willpower; too many, in desperation, have resorted to dangerous drugs.

A physician friend of mine, watching how I had taken control of my weight problem, suggested I share my discovery with others in a short, easy-to-read-book. He wrote the foreword for my book. I was not seeking Dr. Shiu's advice as to the soundness of my weight-loss theories but simply explaining to him that I had lost 22 pounds in 22 weeks without drugs and without additional exercise, low-fat food, a fitness guru, a nutritionist, or any significant exercise of willpower. Before I share with you what I told Dr. Shiu (and have since shared with other friends and associates who are getting the same immediate results that I did) please understand that you and I are truly alike. Otherwise, you may have difficulty believing that you can benefit from my discovery.

I didn't start out with a "weight problem." Like a lot

of people, I had no concerns about my weight for the first 20-plus years of my life. At times and for short periods, I weighed slightly more than average; but spurts of normal, rapid growth quickly eliminated any significant excess weight. I did participate in school sports, but I was not naturally gifted in that way, and had to work hard to improve and to keep myself motivated. And back then, as now, I enjoyed lying around toward the end of the day, and—I admit it—I probably watched and still watch more television than I should.

By my early 20s I had gained noticeable weight, I went up about 20–25 pounds and stayed there for the next decade. All along I was told, and I believed, that my metabolism was slowing down with age, and that this was unfortunate. On top of that, it was assumed that my "ferocious" appetite as a teenager was a big part of my weight problem. Between the pressures of working and going to school I did not have the energy to think all this through and tackle the problem. So I accepted weight gain as just another part of adulthood. After all, I had already experienced numerous physical changes since boyhood. Weight seemed like just one more.

By the time I was 35, I had grown accustomed to not being able to find clothes that fit well, avoiding horizontal stripes, and getting people to concentrate on my great conversational skills at parties. For me, fun social events no longer included dancing, playing sports, or physical activity in general. My loving, caring wife brought home various "diet plans" from associating at work with other loving, caring wives. Together we made several serious attempts at weight-loss programs. Scarsdale was one; we ate a lot of protein, enough over a period of 2 months to simulate weight loss. What actually happened is that we each managed to lose enough liquid to appear nearly 10 pounds lighter for a couple of weekends. But eventually we got thirsty, and the weight came right back. We tried aerobics diligently on tapes at home, but one chocolate donut will run a human at full speed for an hour, and we experienced little if any weight loss. (You can only burn about 450–500 calories max, if you really push it and you're really lucky!) If we'd had any furniture at the time, I'm sure we'd have broken it all—between her 20 and my 40 pounds extra—trying to dance off a combined 216,000 calories to rock-and-roll music in our little house.

Then there were the recipes! Chicken dishes without chicken; butter that isn't real butter; sugar that is sugar-free. The food was awful, and we developed such a dislike for each other's cooking, it drove us to restaurant eating. Our preference for restaurant food has lasted to this day. Does any of this sound familiar? How about diet drinks, light beer, and fat-free, sugar-free desserts? And what about stealing food from your kid's plate at McDonald's? You say you never go that far? Well, we did! Oh, and I almost forgot, you haven't lived until you try to leave a small bite on every plate to train yourself to control your eating.

During all this, needless to say, my blood pressure increased. That concerned me, and I made some efforts to switch to drinking rather than eating. Anyway, to make a long story short, I ended up 60 pounds overweight, over-tired, over-stressed, and over forty. Sound like heart attack material to you? Well, it did to me too, and I decided to walk it off. My wife and I started walking, first 30 minutes a day, then an hour, finally an hour and a half. I enjoyed the long discussions and the chance to review our friendship and talk about the kids. But although there was short-term improvement in my blood pressure, the weight did not go down, and we were in fact growing more and more fatigued from carrying the extra load on our increasingly longer walks.

It's important that you understand: I'm not a nutritionist, or a doctor. I don't have an exercise show or a health studio, and I'm not a self-help therapist. I am not one of them; I am one of us. We are the people who are overweight, the ones who have been used to create a trillion-dollar industry in which "skinny" people (the ones I mentioned earlier) torture us by making us eat terrible things, abuse us physically with exercise at our own expense, and then laugh at us when we fail—or worse—provide us with potentially lethal drugs.

Well the weight (pun intended) is over! I knew it the first day I tried it, and so will you. Pardon me for being silly, but you may want to laugh and jump up and down and yell. You will definitely feel it from your head to your toes, and you will at least want to smile. If you're like me, you'll also be afraid to smile, because you'll have a hard time believing your battle with weight is finally over, but it truly is. In time you'll begin to wonder why you had to struggle for so long. The torture, the insanity, the pure waste! I assure you that you will settle quickly into this new way, and the craziness of your past will be forgotten.

The Engineer's Diet is, in fact, really not a diet at all. You don't need to redirect your food choices or punish yourself. All you need to do is quit doing something you never had any business doing. Eating lunch. Yes, it's that simple: except on holidays and rare occasions, you never should have been eating lunch—particularly as an adult. You didn't ever need it.

All it took for me to solve my weight problem was to fall back on what I know best. Using engineering principles, I was able to devise a straightforward and highly effective plan for proper eating. It boils down to eating one-third less food and eating it on a regular schedule. It was only after the fact that I found I had merely re-discovered what's already been known in some parts of the world for a long time. It seems that three regular meals a day is relatively new in history and even today is not a universal practice. From now on, those of you who go on The Engineer's Diet won't be eating at midday: not fruit, not cottage cheese, not the salad bar, not "nuthin."

Part One

Enough Already!

I DO!

Is Your Body Crying?

MOST PARENTS KNOW THAT new infants cry to be fed. I was fortunate in that my wife gave birth to twins, enabling me to make a great discovery about crying babies and food. After several weeks of being awakened at all hours, by one first, then the other, and watching my wife feed them each time, I was exhausted. But I knew there was another way, and I told my wife it was more likely that the house would burn down than that the kids would starve to death, no matter what we did.

My solution was to have my wife start feeding the twins at 6-hour intervals, whether they cried or not. Within 2 days, not only was I better rested, but the kids and my wife were more content. It was obvious the twins had been crying because they thought they had to. Once we put them on a regular schedule, they no longer had to perform exercises in anxiety and desperation to be able to eat.

Have you been making your body cry to get fed? If so, how would you know it? Keep in mind that your body is not an entity, or an object, but a collection of organs that work together as a team and rely on one another for survival. If you're not remembering to eat, your body's team can put together some pretty amazing activities to get your attention. What about that restless feeling that interrupts your concentration at work every 20 minutes? How about that headache that suddenly comes from out of nowhere? Or the mild dizziness, which hits you when you stand up? Or the gurgling in your stomach that's loud enough to be heard by others in the room. Maybe you've had sweaty palms, or difficulty catching your breath, or even heart palpitations. When people tell me they can't skip lunch because not eating for a whole day makes them nervous or sick, I always know what to ask: "Would you remember to stop and eat if you didn't get all these symptoms?" Most of them admit they would not. I can then point out that their body is being forced to make them anxious and uncomfortable in order to get their attention.

Maybe you're like some of us with lives so busy we never stop running until we experience anxiety or physical discomfort and finally remember to eat? Or maybe you think you ought to be munching "a little something all day long" because you know you'll feel "sick" if you don't. If you're like me, you've always thought you're the only one who really knows when your body should eat. But, think about it: if you aren't eating on a regular schedule, the only way your body has of getting fed is to put itself on alert. So don't be surprised if you feel out of sorts, a little sick, nervous, or "hungry" quite a bit of the time.

Remember, if your body is crying, it's surely *not* because you're starving it to death. You wouldn't be reading this book unless you had food to spare. Within a couple of days, the people who started The Engineer's Diet reported lowered anxiety and hunger symptoms. You can read some of their personal stories in this book.

It wasn't fun for your parents when you were a baby crying to be fed, or for you either if your child did the same. And it's not fun for your family and friends to watch you suffer over your weight as an adult. When your body cries, you need to listen. Put yourself on a regular schedule of eating twice a day; nobody else is going to do it for you. When you start The Engineer's Diet, I guarantee you will become less nervous and more content, just like our twins did. Your body will appreciate the comfort of knowing it will be fed and not having to scare you into eating.

Why Not "Do Lunch"?

THERE ARE GOOD REASONS FOR not eating lunch. First, we're overweight and don't need it. Second, it puts an unnecessary strain on the body. And third, it's immoral to deprive others of food that they might need.

Let's examine the first reason. It's pretty much a matter of agreement that the U.S. is—to a large extent! —a nation of overweight citizens. As of the 1990s, according to the National Institutes of Health, there were 58,000,000 people who reported themselves as overweight. That means having a BMI of 27 if you're a woman, and 28 if you're a man. For you physicists, BMI (body mass index) is found by taking your weight in kilograms divided by your height in meters squared. For those of you who are not mathematical geniuses, it means you'd like to lose at least 20 pounds or the equivalent of two dress or pant sizes. Two things to consider right off: First, those figures represent an underreporting problem, and second, the numbers have continued to climb during the last 8 years.* If we work with a conservative average overweight figure, it

*As this book was going to press, the National Institutes of Health released a 1998 report stating that 97,000,000 adult Americans are now classified as overweight. "Millions More Are Fat," *San Francisco Chronicle*, June 4, 1998.

would be at least 20 pounds. That's based on an overweight woman who wants to weigh about 130 and actually weighs about 150, and a guy who wishes he were 180 and weighs in at 200. Okay:

20 pounds x 58,000,000 people
= 1,160,000,000 pounds
or 580,000 tons of extra weight.

To assist you in visualizing the immensity of this lump of fat, it is equivalent to the weight of 4,000 full-sized, adult blue whales. Of course, the whales have the advantage of water buoyancy in carrying this load, whereas you and I don't. Keep in mind that this is the *conservative* total of how much America is over mass. That alone ought to be reason enough for you to start skipping lunch. And don't worry, there is no danger of starving. In contrast to what you may have read or heard, you cannot starve to death when you are fat. The stored energy your body has in reserve is more than enough to sustain you through the day.

I know that "three squares a day" is the familiar tune you grew up with. So how is it that we all get so round from eating square meals? From an engineering point of view, there are some serious mechanical problems with the three-meal-a-day approach. And these problems can translate into health problems for you. For one thing, eating shortly after the last time you ate does not give your body's digestive system time to "clean up the mess," put things away, and rest before the next meal. This cleanup process is crucial to your survival. Medical experts talk about the need to dispose of free radicals, which are the cause of life-threatening cancers. Does that ring a bell? Sure enough, it's a "chow time" bell. Just look at all the come-ons about the foods you must add to your diet and what supplements you must buy to "cut down on free radicals." Doesn't it seem strange? At the same time as we're all worried about eating too much, we keep adding more and more to our diet to clean the body out. Once again, the medical experts have completely missed the obvious. It is the activity of eating that increases production of free radicals in the first place, and overeating makes this problem even worse.

It seems rare that either doctors or nutritionists state the simple truth that eating less has a positive effect on your health as well as on your longevity. You can read about it in the newspaper. The *New York Times* carried a 1997 story about an experiment with mice; when calories were kept low, their life span increased by 33 percent. The same article reported that calorie restriction in monkeys reduced risk of heart disease and stroke by 30 percent. Pretty straightforward, isn't it? Here's the nutty part! While the researchers agree that calorie restriction works (especially for the guy who's been eating less for two years and feels great) they caution against withholding food from humans. Why? Because, they say, most people can't stick to a diet. That's logic for you!

In general we Americans are quite affluent in that we can get hold of nearly any kind of food in the world to put on our plates. But even though we're too heavy, we're still a vast improvement over the folks in Europe nearly 1,500 years ago who had nowhere near our degree of affluence. I'm referring to Rome, a society that, among other excesses, was apparently able to eat itself to death. They consumed such goodies as pies made of songbird tongues; casseroles of peacock and pheasant brains; parrot-fish livers and entrails of lampreys; and whole pigs served up in cow's wombs. And they ate continuously. For certain, we're doing better than they were—few of us weigh the 500 plus pounds that were all too common in Rome's heyday. But it's not because of willpower; it's because we don't serve fifteen-course meals. So how come we still have such a weight problem? It's simple! We are still serving ourselves an extra course we don't need—the midday meal.

But wait, you say, if it's all about consuming less, why not just eat smaller portions? The answer is that it's really all about efficiency. It takes less physical exertion to digest 1,000 calories in a single meal than it does to digest the same number of calories in ten meals. Why? Because the key to managing the body's food supply is energy conservation. Think in terms of how your digestive equipment works each time food enters the system. When you eat, your body gets ready to fire up the system in order to "batch-process" the food. In other words, when it starts digesting, it's prepared to go ahead and digest a full batch. You wouldn't start up a cement mixer to make a spoonful of concrete!

Partial meals, therefore—long the slogan of obsessive health gurus—are extremely inefficient. The idea is that you'll eat low-fat food all day, repeatedly fire up your digestive system, and lose weight. But the truth is, when you eat several small meals a day, you're not only driving up your caloric need, you're also wearing out your machinery! The result is neither effective nor prolonged weight management; it's wasteful.

Trust your body. You'll discover that its natural efficiency doesn't want you to eat half a meal and walk away from the table. Ever noticed that it's actually not comfortable to do that? If you're honest about how it feels? In fact, I would argue that eating partial servings requires constant vigilance—to the point of preoccupation with food—making it more difficult to eat less than if you hadn't started a meal in the first place. When I find myself starting to eat at all in a midday situation, I just go ahead and eat a full lunch. I don't need the frustration of fighting my body's natural inclination to start and run a full batch. To do any less would feel like "walking off the job."

Here's an experiment you can do at home. Next time you feed your dog, let him get about three or four bites into it. Then walk off with his dish. Notice the look on his face; that's what you look like when you try to leave food on your plate. Over time, your relationship with yourself won't even be as good as your relationship with your pet, if you do this on a consistent basis. And don't be surprised if your dog attempts to finish his meal by eating your hand!

I'M TAKING THIS AWAY FOR YOUR OWN GOOD, BOY.

Fido
GRRRRRRR

Fido

So let's assume you accept as fact that you can let go of some food by skipping a meal. You might be surprised to learn that such a practice is not new. In the *Talmud,* the Jewish guide to proper living written over 2,000 years ago, we read that "as a rule, the mass of people partook of two meals a day." And in India, records from early in the first millennium show that the norm was two meals a day, morning and night. By the way, many thousands of Hindus still follow this practice.

What about skipping breakfast and eating just lunch and dinner? Here's another proverb from the *Talmud:* "Sixty runners ran but did not overtake the man who *breakfasted* early." Sound familiar? Your 90-year-old grandmother probably warned you many times about eating a good breakfast if you wanted to have a good day. Speaking of good days, have you ever noticed that most of the people who don't eat breakfast have some other rather peculiar habits as well? What about drinking a pot of coffee along with lighting up a cigarette? These same folks usually stay up most of the night and sleep in. Most of us would agree that people who don't eat breakfast have trouble getting going in the morning. This makes perfect sense to an engineer—it's a simple matter of not fueling the system. And there's another problem with lunch and dinner as your two meals. Your daily cycle is divided between an 18-hour stretch—from dinner to lunch—with no food, and a 6-hour stretch—from lunch to dinner—in which you eat twice. The 18-hour period is dangerously close to fasting, and the 6-hour period is too short for the cleanup process.

But wouldn't it be better to skip dinner? We've all

heard about how bad it is to eat just before bedtime. They say your metabolism is "too slow" after dark and you should never go to bed on a full stomach because "the food just sits there all night." Well, consider the source. As I pointed out, the gurus of weight-loss programs don't seem to mind obsessing over the details of overly complicated, often unsuccessful eating programs. Many of them are also "fitness" teachers who depend for their livelihood on maintaining a high metabolism. For most of us, the day is done; the evening mealtime is often sociable and can put you in a contented and comfortable state of mind for sleep. What sleepiness you may feel after eating a big dinner is highly inappropriate during the day at the office, but a benefit as you prepare for a night's rest. Another strong argument for not skipping dinner is the fact that again there is an 18-hour stretch between lunch and the next morning's breakfast—taking you into the fasting zone and certainly not using your body most efficiently.

The idea behind putting your body on a regular eating schedule is to give it optimum conditions in which to function well. Skipping either breakfast or dinner won't accomplish this goal because of unequal time gaps occurring between meals. The best choice is obvious: Eliminating lunch gives you the simplest, most regular, and ideal schedule; it requires no significant change in your daily life and gives you the easiest way of all to reduce your caloric consumption.

Now to the issue of morality. I believe I've mentioned the fact that we Americans eat more than we need to. Let's consider this fact seriously alongside another statement: one-third of the world's population is on the brink of starvation while we are overconsuming by the same amount. Is this appropriate in an enlightened age? Remember when you were a kid; your parents would tell you to eat everything on your plate, then try to shame you into it by talking about kids starving in Africa? If you were like me, you answered logically by saying, "Well, then, send them my dinner!" I realize that most of us are not in a position to redirect our personal food, but we might want to take a look at what a grassroots type movement could actually do to help. "Sending our lunch to Africa" might not be an impossible idea after all. A dramatic downward trend in lunch-eating in this country could ultimately force food manufacturers to seek new markets abroad, and maybe food would reach those who need it more than we do.

There are plenty more good reasons for switching to The Engineer's Diet, the No-Lunch Program, or whatever you want to call it. But before I share those with you, I want to cover a bit more about efficiency, nutrition, exercise, and other matters. It won't be tedious or lengthy reading; the idea is so simple. And a lighthearted book seems suitable when the subject is how to become lighter.

Is there a real possibility that American adults could get behind a no-lunch program? Who knows? But of the more than ten individuals I mentioned it to, seven jumped right in and all began averaging about a pound per week of weight loss—not to mention immediate improvements in energy levels, sleeping, and other side benefits that are talked about in the Personal Stories at the end of the book.

Lest you think I have my head in the sand, I've given some thought to the response of the lunch industry—one of the biggest consumer businesses in the U.S. I'm not an investment counselor, but my tip would be to get out of fast-food lunch franchises and get into coffee shops! Which brings me to the subject of coffee breaks. Just because you give up eating at noon, it doesn't mean you can't enjoy the lunch hour. I usually go to lunch and visit with associates and friends, and I have coffee or a soda. The fact that I don't do lunch does not mean I avoid people who do. The only difference between us is that I don't eat—and I don't even snack because that would set a digestive cycle in motion and make me feel inclined to eat a full meal.

Do Kids Eat Rocks?

ONE OF MY RARE PLEASURES while finishing my engineering degree was going to the kids' science fair at the campus gym. I remember one time in particular when I noticed the different competency levels among children of the same age and even from the same class. Side by side were two projects, both by third-graders, but demonstrating a huge difference in scientific comprehension. One was an improvement in the cross-section of the wing design of a Lufthansa jet. The report showed in considerable detail how fuel efficiency would be improved and length of the runway could be reduced.

Right next to that entry was a poster reading: "Do Kids Eat Rocks?" As I recall, this experiment combined hypothetical and empirical work. The report went something like this: Sally said no; Steven said yes; and Billy ate one. Obviously, as I said to my wife, one of these kids had help from the parents. (Of course, if you think about it, we don't actually know which!) My immediate concern was for Billy's eagerness to support science. Notwithstanding, there are some interesting questions raised by the rocks project.

For example: Suppose we actually do "eat rocks," as suggested by the fact that our bodies require many rock-based minerals to survive. Suppose our bodies have very particular ways of indicating preferences for certain foods. And suppose the entire system has evolutionary logic behind it? Without getting too much into hard science, I'll just say to those interested that there's new and relevant scientific literature on re-understanding Darwin's theory and a surge of interest in the exploration of human development in terms of evolutionary function.

Here is the way I generally think about it: For early humans, fruit provided an energy source both high-level and short-term and coincidentally just what was needed for climbing trees to gather the fruit. Grains gave them carbohydrates in a form that burned more slowly than fruit, thus supplying the human machine with energy for the extended task of gathering grain. Protein from meat provided a low-energy food source, building muscle strength sufficient to overcome another animal. Moreover, meat protein burned at a rate low enough that the body could wait patiently for fish or animal prey. Along these same lines, note that honeybees and hummingbirds live on the highest "fuel source" of all—pure sugar; their food-gathering demands wing-speed for hovering in mid-flight over flowers.

So one question is: What should you eat for your lifestyle? For a long time, I told myself I had the active lifestyle of a grain gatherer, so I ate primarily carbohydrates. On that diet I believed I could outperform my friends and neighbors (who, of course, did not have my special nutritional knowledge!) and end up with all the "maids and money." But, I had forgotten two important caveats. First, my lifestyle was far more varied than I admitted; and second, I wasn't really counting on the girth-developing power of one of the most obvious carbohydrates—alcohol. Except for those two drawbacks I have little doubt that my youthful optimism grounded in my basic evolutionary understanding of the human body would have prevailed. More recently I have come to realize that, with the exception of sugar and alcohol, most of the foods I crave prove to be foods that in fact I need.

Examine your lifestyle and see how it is connected to what you eat. Nutritionists may try to convince you that your body is not to be trusted. Nonsense. Here's something to think about: In earlier times, why was scurvy a problem only onboard ship? Why didn't it show up in places where people could get citrus or in people who had a preference for eating only meat? In other words, why do people suffer vitamin deficiencies only in places where the appropriate food sources are not available? The reason is that your body naturally craves foods that supply your needs, even when those foods are difficult to get. All right, then; most of us who are reasonably healthy and live in developed countries, even if we're overweight, should probably follow our instincts as to what to eat. It's the frequency of consumption that needs regulating.

However, we must remember that our "instinctive" desires can be trusted only with respect to natural foods—that is, foods available in almost the state in which they are found in nature. Refined sugar and alcohol do not fall into this general category and probably should not exceed 5 percent of our intake. Fructose is sugar as it is found in nature; that is, in fruits and vegetables. It is an enticement for us to get our vitamin C. But in other foods, in unnatural forms, sugar really is a "sucker punch" to the system. It will do the same job of attracting us toward food, but likely not in the direction we need. If you think artificial sweeteners are better, just remember that the "sweet factor" is the enticer whether it's caloric or not. I'm not saying we should be afraid of all synthesized sweeteners, or all synthesized food either. Just realize that these can affect the body's inclinations to crave what it needs most.

Here's another point about alcohol in conjunction with the body's natural inclinations: Its consumption can definitely dull the sensitivity in your stomach. With alcohol, overeating is easy, because your body doesn't readily notice the discomfort of being too full. In fact, most of us would have difficulty eating huge amounts of food unless we used something like alcohol to kill the pain of stretching the stomach. Why do you suppose alcohol and gluttony are so closely related throughout history? Of course, there are those rare folks who have managed through daytime competitive eating without alcohol to stretch their stomachs to huge capacity. But now we are learning from The Engineer's Diet that any oversized consumptive capacity disappears rapidly with a routine of eating twice a day.

More about trusting your body. Doctors have long observed that small children pick out what their bodies need from a variety of foods. How many parents have raced to the pediatrician worried because "Bobby won't eat his broccoli," or "Sally takes the turkey out of her sandwich." Only to have the doctor say the kid's fine and healthy and you shouldn't worry, because kids manage to get what they need. With time, for some unexplained reason, this talent supposedly disappears. Older children and adults apparently lack any instincts for what and how to eat! Now why would that be?

A frequent problem with physicians is that they aren't rigorously trained in the scientific method, so they have little or no concept of a proper control study. They look for common factors rather than for differences. Here's what I mean: Suppose you took a group of doctors into a junkyard full of cars to determine what causes wrecks. Using a standard medical approach, the majority might say it was dead batteries. Instead of paying attention to the differences in damage sustained, many of them would deduce that because all the cars had dead batteries, dead batteries caused the wrecks. Unfortunately this same kind of thinking has permeated the health care community to such an extent that certain food groups, which are in fact quite good for us, have essentially been banned by the medical profession. I'm referring to red meat, eggs, and dairy products—the top three foods on the current diet "hit" list. Here's the reasoning from the doctors: Overweight people tend to have serious health problems such as heart disease, high blood pressure, diabetes, and various cancers. Overweight people are people who overeat. And people who overeat are also those who eat red meat, eggs, and dairy products. Ergo: Red meat, eggs and dairy products *cause* serious health problems. The case is clear: "Dead batteries cause wrecks." The fact that there are cars with dead batteries that were never in a wreck, or that there are numerous cultures with diets high in fat and protein and very low rates of these diseases, does not shake the faith of the modern physician or nutritionist. The consensus is that red meat, eggs, and dairy all gotta go.

As an engineer, I must ask some obvious questions about these assumptions. For example, what if the only fat really dangerous to humans is human fat? If that were the case, obese or seriously overweight people would still have medical problems, not necessarily because they *eat* fat, but because they *are* fat. There's a big difference. But this more logical conclusion could work even for the doctors because people who were fat could still be diseased and people who weren't could still be healthy—which is in fact what the medical evidence indicates.

I can't stress too much that you should probably eat what your body tells you to—assuming it is not constantly telling you to eat just sugar. Remember, for good weight management, it's not a matter of what you eat as much as it is number of calories—units of energy—that you take in vs. how much energy you put out. That is why there's so much emphasis these days on exercise. Unfortunately, those advocating increased exercise without calorie cutting are not only *not helping* with the weight problem, they are creating some new problems in connection with metabolism rates. As are the advocates of eating smaller meals back-to-back. The fact is that continuous exercise or continuous eating is known to raise metabolism—an effect, by the way, that almost all modern diets have tried to exploit.

I'll go into the facts about increased metabolism later in the book, but I want to make a particular point here. Continuous raising of your metabolic rate as you grow older is not a desirable goal. In fact, the human body, including its aging process, was designed perfectly for maximum efficiency and that, as I mentioned earlier, includes conservation. In the section on timing, I'll say more about this. And when I talk about exercise, I'll explain the metabolic aspect of optimum health. But the bottom line on which nearly everyone agrees is that if you sit on your duff all day and eat too much, you'll be fat. On the other hand, if you're skipping lunch and moving around in normal fashion, you needn't be afraid to eat and enjoy all kinds of food—fat, protein, starch, and dairy. And furthermore, you can eat until you're not hungry. Imagine that!

Weight Management: Engineering-Wise

AT THIS POINT, YOU MAY BE telling yourself that the No-Lunch Plan is too, too simple. So it can't possibly have merit. After all, why have doctors, nutritionists, scientists—even whole medical schools—spent so much time and money researching weight and weight management? They must know something! And what about the food industry? Thousands of new food products have been developed, manufactured, tested for FDA approval, and recommended by the medical establishment. Isn't this science? Shouldn't we pay attention?

You want a simple answer? Okay, the answer is *no*. You don't have to pay attention to most of what's been discovered in modern times. The current media craze with dieting, nutrition, and body image is just that—a media craze. Remember that throughout human history, we've been challenged to find good sources of calories that were both healthy and pleasant to eat. Usually, however, those challenges have been natural and often temporary—caused by changes in elemental conditions of the habitat. Then too, there are frequent human-made obstacles to healthy diets all over the globe—government policies, war, population displacement, and so on. I simply don't believe we need to panic that the planet is running out of safe, nutritious food. A lot of what is shaping our current thinking is politics and, as such, will change frequently.

So I suggest you take a break from immersing yourself in the subject of dieting and health. I can suggest one such area, however, that is of interest—that's the latest research on longevity in connection with lower calorie consumption. Also, check into recent findings that low-fat and no-fat diets may be linked to increased risk of strokes.

A Solution-Oriented Approach

An engineering approach to weight management is going to differ significantly from the health care approach, for obvious reasons. For one thing, some of the best aspects of health care (when it's functioning well) lie in its purpose of trying to educate patients, thus bringing about change in overall health through modification of patient behavior. Probably the biggest examples of this are efforts to raise public awareness about the significant effects of personal hygiene, about proper management of food to avoid contamination and spoilage, and about ways to moderate unhealthy behavior such as overeating, drinking too much alcohol, smoking, and so on. And, of course, mainstream medicine and surgery also provide standardized relief for the sick and injured.

As an engineer, however, I would argue that the greatest improvements in health and longevity for populations at large have been brought about more directly by civil engineering. I'm referring, for example, to the provision of clean water and the design and function of proper waste disposal. There's solid logic as to why engineering solutions have more immediate and more widespread success than medical solutions. Engineering solutions, by the very nature of the discipline, are not aimed at slow and difficult (if not impossible) reeducation of people. On the contrary, they are geared to solutions making low demands on the individual.

Let me give you an example, perhaps a bit out of the ordinary, but it should make the point. Say, there's a man who's decided he wants to go live underwater. What would a psychiatrist do? Start questioning the man's motives, for sure. And a consulted physician would certainly want to do tests to rule out a neurological disorder. What would I as an engineer do? I'd want to know how far down the man wants to go, and how big a space he'd need. By the same token, a vast majority of people in the U.S., and probably in other developed countries, would like to lose some pounds and maintain themselves in the weight range normally considered healthy for their height and build. They don't need a complicated, drawn-out program! They just want to get on with it. They need an engineering approach—simple and straightforward: a standard industry approach—a three-phase project consisting of Analysis, Design, and Implementation. Let's break these down as an engineer would:

Phase 1. Analysis

People become overweight because they consume more energy than they put out.

I WANT TO LIVE UNDER WATER!
ACME ENGINEERING
HOW DEEP?

Phase 2. Design

(These would usually be set forth as constraints or limits.)

1. Humans like to eat.
2. Once a human starts eating, he or she feels a compulsion to fill up.
3. Humans have a complex selection system including attraction to foods their bodies need.
4. Humans have great difficulty utilizing willpower for extended periods.
5. Humans, especially overweight ones, are often highly attracted to sugar and alcohol.

Here, in the design phase, is where the engineering approach differs significantly from the medical approach. Accepting design constraints is the objective of engineering. Given that no design is perfect, optimization of old behavior and minimization of new behavior are the solution most likely to be accepted with the least resistance and to be sustained over the longest period of time.

The new plan needs to accommodate rather than to fight the constraints—in this case human desires and instincts—and to minimize the demand for behavior change, which draws heavily on willpower. There is a problem with trying to educate individuals to a higher level of urgency calling for greater exercises of willpower. Like most plans predicated on human fear, the "scare tactics" of the medical experts fail because the body cannot sustain a constant level of fear. As the fear dissipates, previous eating habits resume. This becomes obvious when you consider how many heart bypass patients regain so much weight within the year following surgery. Not to mention the number that have to get another bypass within 5 years of the first one!

With all this in mind, we can develop a pretty successful plan within the constraints: Based on constraints 1 and 2, the subject should be allowed to eat until he/she is reasonably full. Number 3 says the subject should choose from a wide variety of foods and rely on desire for selection. Constraint 4 is related to regularity and frequency. Constraint 5 is problematic in that there are "tricks" that certain types of food can play on the body, so there is more to consider here. More about sugar and alcohol in a later chapter.

In essence we're saying the subject should fill up on whatever foods he or she "craves." Then we decide about frequency to help solve the "willpower" problem. Remember our twins? It was the guaranteed schedule that quieted them down, making it unnecessary for them to get agitated and hungry. For us adults, that translates to not having to "exercise" willpower. Okay, what are the choices? To be eating at consistently regular intervals, you'd have to eat every so many hours round the clock, right? Every 4 hours, 6 hours, 8 hours, etc. Let's look. Supposing you decide 4 times a day—every 6 hours. That might seem to make sense because you figure you would keep doing what you're already doing and just add another meal, right? Well it isn't really such a great idea. For the 12-hour gap between dinner and breakfast that you need to cut in half, you'll have to start setting your alarm for midnight supper! You could try this, but you might not be too happy getting out of bed to prepare and eat a full meal on a nightly basis. Besides, have you looked in the mirror? If you needed to add another daily meal to your life, you wouldn't be reading this book, would you?

What about every 8 hours? 6 a.m., 2 p.m., and 10 p.m. There are places in the world where that's standard—Spain is one. They also start the workday late, shut business down for three hours in the afternoon, and end the workday around 9 p.m. Seems like in this country you'd have to somewhat rearrange your business and social life. Also, you would still be consuming at least one-third more calories than you need, unless you measure portions to reduce calories. So four and three times a day may not work too well. How about once a day, you might ask—to make it simplest of all? The truth is, you probably can't eat enough in one stop to meet your body's needs for 24 hours. That's pretty much known and agreed on.

So, we can see the logic and the convenience of 12-hour intervals. Once upon rising (assuming you're a nighttime sleeper) and once before going to bed. Try considering the extra fat on your body as the food you've been saving for a rainy day. Well, guess what—it's raining!

Another way to think about the advisability of eating once every 12 hours: When you eat (fill up), your stomach (gas tank) gets heavier; when you're on empty, you're lighter. Make sense? So engineering principles would dictate that you fill your stomach when it is near empty, insuring a minimum load to carry and thus optimizing the efficiency of your body (vehicle). Keeping the tank full at all times would require numerous interruptions (fuel stops), and also add carrying load to the vehicle. Hauling around that midday meal taxes your body by making it work harder to handle the heavier load. Of course, you could eat almost continuously, and you should, if you are anticipating a famine; but my assumption is that you are usually within reasonable distance of a grocery store.

Phase 3. Implementation

Eat breakfast, skip lunch, eat dinner. On special occasions, feel free to vary the program.

Maintenance Plan

(Ongoing for the rest of your life.)

Every good plan contains instructions for maintenance. Here's how to maintain your new eating program. Notice when you're eating; if it's not breakfast or dinner, then you're not on the program. If you're not on it, then get back on. If you are wondering how to handle your emotions about having temporarily gotten off the two-meal program, I can only say that "guilt" is not an engineering term.

Measuring Is Not Important

No other aspect of dieting has received more attention than the process of measuring portions. For most people constant preoccupation seems to produce more anxiety and guilt than it does measurable weight loss. This is because the medical model of weight control is a complex scheme in which mental and physiological health is viewed as highly influenced by environmental and emotional responses. Doctors try to improve patients' ability to maintain controlled eating by raising awareness of the dangers of obesity. As I said earlier, this means they try to scare the hell out of us. They may have to prescribe hunger-suppressant drugs (because fear is an inadequate behavior modifier) and then tranquilizers, because the fear affects our ability to function in other ways. It's not surprising that weight-troubled patients often end up under the care of a psychiatrist or therapist! The truth is that the success rate is pretty dismal for significant and permanent weight loss using the medical approach. The medical profession turns away from this failure, claiming that obesity is an "ongoing dilemma." And for business-minded doctors, long-term "disease" generates lucrative side effects—poor health conditions calling for such procedures as bypass surgery, liposuction, wiring the jaws shut, or sewing up portions of the intestines.

Nutritionists are still another matter. They maybe wouldn't even have jobs were it not for the failure of people to control portions. At the most, the nutritionist's job with an overweight person is to give a 5-minute description on basic healthy eating. That's it. But, once it is apparent that almost no one can control portions, a whole new world of lifelong careers opens up. There are fat levels and sugar levels to calculate; percentages of minerals and elements to organize into unreadable tables; lists of scientific names for all of the known and unknown foods; by genus, species, phylum; papers and books to write; and talk shows on which to appear. All in an effort to control weight in persons said to be suffering from "uncontrolled eating." Just remember this:

Nutritionists have nothing to gain from our long-term ability to solve our weight problem—in fact, our success probably threatens theirs. Engineers, on the other hand, have nothing to gain either from our success or our failure with weight control. They aren't in any danger of losing livelihood either way. They can view the entire issue simply as an engineering problem—it's probably got a solution; figure it out and fix it.

One of the first questions people ask is: How long will it take to get used to skipping lunch? Not that long. You're already going from dinner to breakfast without food—that is, unless you're a midnight snacker as well. Your body is actually used to waiting somewhere around 12 hours. If you're not fainting on your way to breakfast, this should tell you that your body is working on stored energy. You'll find that you can actually move around in a number of ways before you start loading in the cereal, eggs, or waffles. In fact when your body stores fat, it tends to store it around the waist and trunk so that you can have maximum use of your legs and arms. Your body knows how to store energy efficiently, both for high and low activity. Here's another way to think about it: Notice how you can continue to drive your car until the gas gauge is actually on empty? Well, your body works the same way; you just have to begin to trust it. The next question is usually: Won't I be so hungry by dinner that I'll overeat? The answer is no. If you stick with this for awhile your stomach will shrink, and you won't be able to. Several people reported that they did panic at the very beginning and tended to load up on dinner at first, but within less than a week, all of them agreed that they had settled comfortably into the plan.

There are, however, some really critical issues to deal with, such as: What do I do with my extra lunch money? This is an important concern, because if you are currently overweight, you'll soon be losing a pound a week and will need to buy smaller clothes.

Maybe you've started to figure out that if you eat this way from now on—and lose about a pound a week—you'll just go on losing. What keeps you from losing too much? It's true, you're wiping out a third of your caloric consumption, so a bit more at breakfast and dinner may be indicated when you get to your desired weight. You might ask why you can't eat a small lunch like a salad or a turkey sandwich. Shouldn't you be able to control yourself once you've achieved your goal? What's the difference if you have 220 calories at lunch or extra calories at breakfast or dinner?

Actually the answer is quite simple. Added calories at breakfast or dinner will leave you quite full, whereas only 220 calories at lunch will "fire up the system," leaving you quite hungry. Remember, if you don't start a "batch-process" at lunch you will not need Herculean willpower to resist the urge to complete it. Also, you are bound to tire of lean turkey or green salad. You may intend to keep a low-calorie routine, but you'll probably start eating something a bit more elaborate every so often. Pretty soon it could be a burger, or a pasta dish, or soup and sandwich two or three times a week. Guaranteed you will gain back at least 10 pounds a year, which means you'll be inclined to start over-exercising; in which case, you'll wear out your body and probably die younger. Instead of all that, why not enjoy a nice cup of tea?

Going back to eating lunch would be nothing more than reentering what is fast becoming America's No. 1 sport—competitive eating. In case you decide to race with the champions, here's what you're up against. Historian Charles Gattey writes that the Roman emperor Claudius the Lame "hardly left the table until he had crammed himself with food and drunk to intoxication. Then he would fall asleep lying upon his back with his mouth open. While in this state, a feather would usually be put down his throat to make him throw up the contents of his stomach so that he could start all over again." I urge you not to play in this league; leave the food to those who want to be immortalized in the Lunch-Timers Hall of Fame.

Let's be honest, if you could have successfully managed your life with self-discipline and paid attention to every morsel you ate, you wouldn't be reading this right now. I'm not saying that controlled and selective eating doesn't have its place. It's perfect for millions of Americans who wish to prolong the weight struggle by working in opposition to their natural body demands. Remember, my approach to eating properly does not involve changing your basic desires for what kind of food and what size portions to eat. The idea here is to accept your biological urges as healthy and efficient and to do what your body wants you to do in the way your body wants you to do it—just not as often.

Does this all seem too easy? It is. And that's just what it should be. Not everything worthwhile has to be difficult. Which reminds me, I have a real problem with the statement, "No pain, no gain." I find it utterly absurd. But the truth is, I too fell victim to it when I was much younger. I worked diligently at being an athlete all through high school. And I bought into the idea that no matter what the goal was or how much natural talent was involved, you should always work harder, even if it hurt. But, I've noticed through the years that lots of dancers and gymnasts who had great "promise" wore themselves out or got injured in training and ended up not making it. To this day, I wonder if I could have had a successful athletic career if I had simply not worked at it so hard. On the other hand, if you still insist on using the maxim, try structuring it differently. When you start skipping lunch, you'll truly have "No pain *and* no gain!"

Timing Is Everything

When you eat, the digestion process, which is both chemical and mechanical, requires a higher rate of metabolism. Metabolic rate is the pace at which your body handles the breakdown and dissemination of whatever you consume. Again think of your body as an engine. This engine's idle changes throughout your life. That's your timing mechanism. Nature starts you out with a high idle; the idea is for you to run around and work up the energy you need to develop your bones, lungs, muscles, and a good, strong frame. Kids "exercise" without much effort because their metabolism is high. As you get older, your metabolism naturally begins to go down. That's because once you've gotten to the age where your body is fully developed, you don't need a high metabolism. From nature's point of view, a high metabolic rate in an adult is wasteful. Kids can get the energy they need for all their activity on half of what most adults eat, just because of their size.

After your body is fully built, to have you trying to burn twice as many calories a day would be poor planning. Nature is a natural conservationist knowing enough not to create a big demand on the environment to provide you constantly with food. And in that conservation mode, whenever you as an adult have an emergency activity that requires a higher level, adrenaline stimulates your body, raises you up to the required level of performance, and then lets you back down again.

You were originally designed to carry on an excellent conservation program. To understand this easily, watch your dog. When it was a puppy, it was boing! boing! boing! everywhere while it was developing. One day you looked around and realized your dog was doing nothing but sleeping. But now when you feed him, as you get the food out of the bag, notice how you are involved in simulating "the hunt"—even if the prey is just coming out of the doggie food bag. The dog gets very excited, jumping around and barking like crazy. Then right after it eats, back to sleep. This animal is conserving—running at a lower idle except when it needs to run higher to secure the necessary calories. Everything is aimed naturally toward conservation. Historically, animals that depleted their resources did not survive. Those with the best conservation equipment persevered. So it is with us humans.

We've seen that the American style of eating three meals a day tends to produce either fat people or people who are skinny but constantly preoccupied with food—and occasionally those who are happy to trade off potential longevity for the extra benefits they derive from competing in athletics. So its logical that it's better for adults to eat twice—once in the morning and once at night. Also, given the wear and tear on your digestive system that eating more often creates, it would also seem obvious that minimizing digestion is advisable.

But not to the nutritionists these days. Many current diets call for constant "snacking" throughout the day to raise your metabolism in an effort to burn fat. Remember, this is not only inefficient for your body, but goes against the body's natural inclination to optimize its resources. But the books go right on recommending that you eat numerous, tiny, low-fat or nonfat portions all day long. They even sometimes call it the "grazing" style.

Okay, let's consider this approach. Examples of champion grazers are all around us; a contented Jersey or Guernsey cow is a beautiful sight to see. And they eat the lowest-fat, greenest diet you can find. Are they really appropriate role models? If your goals include producing butter, cheese, and cream, or hundreds of burger patties over a 7-year life span, then the no-fat grazing diet is just right for you! Of course a crafty nutritionist might point out that all-day snacking is natural to monkeys and baboons as well. As an engineer, I would have to agree that tree animals do eat fruit all day—especially when it's in season—in order to be prepared for scarce times. So if your goal is to gain weight for those "hard" times in the future, then go for it! On the other hand if you want to maintain a healthy robust lifestyle in the present, then eat a varied selection of foods about every 12 hours, at breakfast and dinner.

HEY MOLLY! EATING THESE NON-FAT, SMALL MEALS ALL DAY LONG I'VE GAINED 1400 POUNDS!
YEAH, ME TOO! I'VE GAINED MORE THAN I NEED TO MAKE MY BUTTER, MILK, AND CHEESE QUOTAS.

The Broader Implications

I LIKE TO THINK THAT ANYTHING we benefit from individually has a reverberating effect on the environment we live in—at least that's the way it should be. And losing weight is no different. If you're like me, you've been buried so long under the stress of how to solve your personal weight problem, you haven't stopped to reflect on the broader implications. Along with the obvious results of feeling better, looking better, lowering your blood pressure and therefore your stress and anxiety, there are still more plusses. Lots of them. Yes, as the mass of America shrinks (I use mass because of the material reality: If we're fat here on Earth, we'll also be fat in space), both micro- and macro-benefits accrue.

Consider this: Are you afraid to fly? Millions of Americans are. Did you know that some of your fear of flying may not be based on the reliability of the plane but on the knowledge that more than a third of the 280 passengers are also overweight? You might wonder how people who build planes could possibly know how fat most of the passengers are; you like everyone else usually lie about your weight. So where do the builders get their passenger-load figures? And if you're like me and run the numbers at lightning speed, you're dead certain that plane isn't going to get off the ground.

Actually there are a few things you need to know.

First, we engineers know that most people lie about weight (because we do too!) and we've designed airplanes to carry *actual* rather than reported weights. Second, as a thinner passenger you'll be less anxious during the flight if you feel better and have low blood pressure. Finally, the truth is planes *can* fly significantly better in terms of speed and fuel efficiency if everyone gets rid of that extra 30 pounds. In other words, skipping lunch will help the plane "skip" off the runway, so give the pilot a break. Besides, you might have noticed, as I have, that the airline food industry does its part to discourage us from eating! Maybe now you'll refuse airline lunch service and ask for water, tea, or coffee instead.

Shoes. Okay, what's that got to do with skipping lunch? Well, the implications from an engineering perspective are mixed. The wear and tear on your shoes come about as a result of friction, a blend of the friction coefficient and normal force. For non-engineers I'll say it more simply: You, as the normal force, wear out your shoes faster if you weigh more. When you walk, you tend to slide your feet more. At less weight, you'll lift your feet higher in a new, bouncy walk. You might find that the soles of your shoes wear out more evenly and have to be re-soled less often, and the whole shoe may last longer. But there is a downside. Once that extra weight is gone, and you start moving a lot more, your shoes could actually wear out quicker! That means more shopping for shoes. Of course, there's a bright side to that too. You'll be giving the shoe industry more business. That boosts the economy and adds a little joy to the life of the shoe seller.

On the subject of personal economy, let me point out that you'll cut your noontime spending in half by not eating lunch. Only half, you say? How so? Well, it would be more, but coffee has gotten to be nearly half the price of a meal these days, so if you're going to do like I do and drink some coffee at your noon break, it will cost you. But, hey, any savings is a plus, right?

Speaking of savings, wardrobe will rapidly loom as an area of great interest and changes as you approach your target weight. Right now you'll have to admit that the stripe has yet to be invented that can hide your substantial caloric endowment—horizontal or vertical. Well, one of the great joys of my new lifestyle is throwing away my old clothes whenever I go down one size. And I mean *throwing* them away. On previous weight-loss programs, I used to save them—sort of hide them away—because I was never quite sure when that lost weight would show up again. Now I confidently march to the trash can with my 42-inch waist jeans because I know if I don't eat lunch, I'll never gain it back again.

On the subject of wardrobe, here's another thought. From an engineer's perspective, clothing is basically packaging, just like transportation is shipping of goods. Given that, I predict that the nation will realize tremendous savings in the "body-packaging industry" just from human "downsizing." For example, let's say there are 250,000,000 Americans and 50 percent of them are significantly overweight. (I'm using a figure that is probably closer to the truth than the statistics will tell you.) That means about 125,000,000 overweight. Let's conservatively assume that their waist size is about 4 inches bigger than it should be. Remember, we're being conservative. Okay, let's calculate the amount of cloth it takes to go around these extra inches:

125,000,000 Individuals x 4 = 500,000,000 inches.
Divide by 12 to get feet. It rounds off to 41,666,666 feet.
Divide by 5,280 to get the miles. You get about 7,900.

Now assume everybody gets about 20 new pairs of pants, shorts, skirts, etc., annually. We end up with 158,000 miles of extra fabric. It's 237,000 miles to the moon, right? If the overweight Americans each lost just those 4 inches, we could stretch the length of fabric we saved more than halfway to the moon! Every year! And that's only counting Americans!

Here's how the savings might look, simply because people who weigh less need less fabric to cover their bodies....Manufacturers might have to buy less from cotton farmers, who might then become more efficient at farming and need less government subsidy, which might in some way lower our taxes. Yeah, right! Anyway, more efficient farming could result in selling cotton to clothing manufacturers at better prices, which might lower the cost of manufacturing. Also, at the factories, less electric energy would be required to turn the smaller amount of fabric into clothing. That might make owners see fit to raise wages. Sure it would! Also, if the manufacturers were able to get more piece goods per fabric inch, they might have more articles of clothing to sell. This would result in more merchandise in the stores, which should bring the prices down, meaning you could better afford the increasingly smaller new clothes you'll be needing. Then, too, there would be less refuse for landfills from the clothes everybody discards when they no longer fit—because people's wardrobes won't be so worn out and will be more suitable for resale clothing outlets, which means more people with less money will be able to get nicer clothes, and we could go on like this forever....

The savings story doesn't end here. You'll also save money shipping yourself about! A 25-pound weight loss will improve the fuel efficiency of your car. In fact, if you're married and can get your spouse to participate in the no-lunch program, you can probably add as much as another 50-pound loss for a total of 75. (I'm assuming your spouse has a worse weight problem than you do or he/she would be reading this book right now on your behalf. Reader surveys indicate that most folks buy a book like this for someone else who needs it!) By the way, notice how easily an engineer can practice psychology? I daresay, it wouldn't work so well the other way around.

Getting back to the numbers—a 75-pound reduction in the family car load would amount to about a 3 percent increase in mileage. With gasoline at $1.20 per gallon, that's almost $.04 a mile (not even including the mileage saved by not driving out to lunch on weekends). Assuming a minimum driving average of 30 miles a day—shorter than most people's probably—you could save at least $438.00 a year. That is the cost of at least one, maybe even two airline tickets! You and your spouse could take a trip. And, as we pointed out earlier, if enough people slim down, the airlines might pass on the savings benefit they get from better fuel efficiency to you—well, maybe I'm getting into a little wishful thinking here! At any rate, get ready for this: Assuming an average driving life of 50 years, you could save $21,900.00. By taking off that extra weight! Okay, let's say you think that close to $22,000.00 over 50 years isn't that much. How about considering it alongside your grocery bill reduced by perhaps one-third, the savings of not having to buy extra-big clothing, the absence of huge doctor's bills from weight-related health problems, even the possible bypass surgery you don't have to have? Not to mention the savings on your water bill due to the reduction in the number of dishes that need to be washed. And how about toilet flushes? There'd be less of those over time, you can be sure. Now, multiply all that saved money by the number of overweight people in this country—and you'll see that the financial consequences of continuing to eat lunch are serious, if not absolutely staggering.

I could write an encyclopedia on the different species of land, air, and water animals that would be spared being taken from their havens and summarily executed for the noonday special! The next time you decide to treat your office partners or your support team to lunch, don't.

Of course, nothing is without its minor problems. Take utility costs, for instance. Unlike transportation, utility costs for public-buildings are affected both negatively and positively by overweight people. In the summer, air conditioning costs in a crowded building will be noticeably higher because of the number of large-mass objects radiating at 98.6 degrees. Also, thermostats have to be kept low because heavier people tend to feel continually overheated. In the southwestern region of the country, this can turn into some significant dollars. On the other hand, in the winter, heavier folks may well contribute to slightly lower heating bills. This may be one instance where the benefits of weight loss may appear to be offset by the liabilities—at least on the surface!

But let's think up a few more obvious conveniences of losing weight—for example, maybe you ride horses.

Well, your horse will be happier. Needless to say, whatever kind of animal you like to mount will definitely appreciate your weighing less! Which reminds me: Dancing could be more fun too. Of course, this brings up another minor problem. You might become more attractive; which in itself could be great. But if you're married and worried that you'll start drawing attention from people other than your spouse, you might want to make plans for putting out some fires. In general, though, the new job of "managing" your fan club has got to be a step up over the old job of managing your weight!

One thing I really like now is not having to wake up in the middle of the night to put the fitted sheet corners back on. At my maximum mass, they were always jumping off the mattress because the bed sagged down so much in the center. Not to mention that a good, king-sized, box spring set can cost more than $3,000. Sagging springs are not just uncomfortable anymore, they're a serious expense!

At this point, I want to focus on a particular body part that you might not think is affected by your weight problem. I'm talking about your teeth. This is pretty basic—not some complicated idea about the effect of food enzymes or acids or other elements involved in passing food through your mouth, and it has nothing to do with the issue of whether you should regulate what foods come in contact with your teeth—with the exception of sugar, that is. No, I'm talking about a simple matter of recognizing that your teeth will wear down faster the more often you use them! I'm not aware of any evidence whatsoever that low-fat foods will protect your teeth, but it is a mechanical certainty that frequent chewing will contribute to wear. Eating twice a day instead of three times has to be beneficial. It's worth at least an investigation by experts in the field, wouldn't you say? While I'm on the subject of things that ought to be investigated, what about the assumption that we need to brush so often? I recently asked a dentist if nature provided any protective coating for the teeth. My thinking was that if nature did, then brushing it off—especially using all those chemicals in toothpaste and mouthwash—would remove the coating and make the teeth vulnerable to faster decay or even decay that might not have occurred at all. The dentist said he had never encountered anything about potentially adverse effects during his medical training, then immediately labeled my hypothesis absurd. I wonder why he did that so fast? Engineers have long known, for example, about the protective benefits of leaving rust on steel beams. I am not recommending here that folks stop brushing. My point is that dental specialists should at least have enough respect for the human body to determine that brushing is not hastening deterioration, and also whether some types of toothbrushes and toothpastes have a less damaging effect than others. After all, it is just possible that the human mouth was a fairly reliable device even before toothpaste and mouthwash!

And speaking of teeth—This last Halloween I didn't eat any lunch but I did eat some of my kids' trick-or-treat candy. Should I be worried? I'm bound to say that any parent who would do such a thing is a creep and a louse, but that would probably include all of us, wouldn't it? So just tell the kids they'll get their turn when they have kids. From a weight maintenance point of view you can relax, because the damage is insignificant. Wipe out lunches from now on and you have eliminated approximately 30,000 candy bars' worth of calories from your diet over the remainder of your life. The five or ten you steal from your kids once a year won't matter. Just remember three simple things: Don't eat lunch; don't make sugar a habit; and don't get caught stealing Halloween candy if you want your kids to respect you in the morning.

Part Two

It's As Easy As 1...2!

The Engineer's Diet How-To

YOU'VE READ QUICKLY THROUGH the first half of the book, and it makes enough sense that you're willing to give it a try. What should you do? Well, it's pretty simple, but I'll make a list for those who like lists and want to feel as if they're doing something substantially different. The fact is, you are making a significant change, but it doesn't take a significant effort. Here's what I did (and still do, for the most part):

A. Get up and eat breakfast.

I eat a bowl of cereal, or a bagel, or both. Eat normal breakfast food and base it on any number of things, but realize that you are best off eating what you like. You probably have some cultural or family history that plays into it—foods you grew up with or special tastes acquired over the years. That's fine. I'm assuming most of you don't eat a plate of cotton candy and drink a quart of beer at breakfast. A few people tell me they eat only fruit. One woman eats eggs, toast, and bacon at least once a week. And one man doesn't hesitate to eat pancakes, waffles, or French toast. It's really up to you what you do.

Early on, one or two beginners complained of a mild stomach discomfort that hit somewhere in the middle part of the day. Sounded to me like an acid stomach. If your body's used to having to digest at more frequent intervals, it's doing the right thing by starting to produce the needed acid at the expected time. I had the sufferers drink grapefruit juice with breakfast every morning—not orange juice, mind you; it isn't the same—and within a week the problem went away. What this does is let the body know that enough acid is already available, and no more need be made. So if you're one of those who has the acid problem, drink grapefruit juice for awhile, and see how you feel. It should take care of it.

B. At midday, take a break and have a drink of something (non-alcoholic).

I usually have a soda or a coffee, often more than one cup. Some folks have tea or sparkling water. If you want accelerated weight loss at the beginning, have a drink that has few or no calories. But remember, this program is a lifelong deal. Think about the rate of weight loss you're after. If you lose only a pound a week—or even only 4 pounds a month—within a year, you'll be down 48 to 52 pounds! Can you even afford to lose that much?

C. Anywhere from about 5 p.m. on, start thinking about dinner.

I usually have a few crackers and some juice while I decide what to eat or while I cook my meal, or just while I'm waiting—if dinner is to be a social occasion. Basically, you can eat what you like at dinner as well. If you want a jump start on the weight loss, limit yourself to around 1,000 calories. "But," you say, "I thought I didn't have to count calories!" Well, you don't. If you eat a normal dinner, and eat until you are comfortably full, you will lose weight at the rate of about one pound a week. Nevertheless, there are those who wish to lose weight faster at the beginning—I was one—and a way to do it is eat a lower number of calories at both breakfast and dinner. But you should be prepared to have to increase calories once you reach your goal weight. Maybe you'll have to add dessert. You know what they say: "Sometimes we just have to make sacrifices!"

D. Repeat A, B, and C for the rest of your life.

A Word about Holidays

In part one, I mentioned that holidays are different. Not only do we in the U.S. equate holidays and special occasions with an increase in eating; it's almost universal that feasting goes with celebrating. There's nothing inherently wrong with this; plenty of evidence exists in ancient texts that food has served not only as physical sustenance, but social and spiritual as well. In the *Talmud,* for example, there's mention of an "extra meal" that can be eaten on the Sabbath. So it would seem there's good historical precedence for eating more on holidays. And that indicates an evolutionary if not natural dimension that should be considered. Remember, the engineering approach recommends changing only what is necessary to achieve the design. Eating an extra meal on the holidays is not significantly going to affect a lifelong style of eating only twice a day. But, if you

EAT BREAKFAST
SKIP LUNCH
EAT DINNER
FEEL FREE TO VARY
—ON—
Special Occasions

do find that over time your daily calendar has begun to look like a festival of exotic, round-the-world holidays, you might want to ask yourself if something else is up!

Eating out on The Engineer's Diet is easy. To start with, you can have any kind of breakfast food you want. At dinner, same thing. Go where you want, eat what you want. And remember, you don't have to have certain-sized portions, or nonfat foods, nor must you leave anything on your plate. If you are one of those who bear the ugly scars of calorie-counting, you will automatically calculate every item on the menu. But be assured that, over time, you will heal, and those rapid-fire additions inside your head will slowly fade away.

Going out to lunch is also fun and trouble-free. Wherever you go, just order the Engineer's Plate. If they don't have it, that's great. That's exactly what you want, with a cup of coffee, or tea, or a soda. If they do happen to have it, then tell them fine, you'll have just the plate, please. And while I'm on the subject of fun.... It's important that you continue to keep your lunch hour on schedule. Go have a cup of something and relax. Although you are used to regarding the noon hour as a trip to the feedlot, in fact, it is a necessary break from the work you have been doing all morning. Whether you knew it or not, your body was telling you to slow down and rest. Some folks talk about the thrill of finding another hour in which to "get stuff done" now that they aren't eating lunch anymore. But I caution you strongly against using your new free time to increase your productivity. The danger is that you will ignore your body's need to slow down and end up putting stress on yourself instead. Eventually this could even lead you back to the table to eat.

Le Bistro
LUNCH SPECIAL!
ENGINEER'S PLATE

"Watching" Your Weight Disappear

Most people want to follow their weight loss, at least at the beginning. If you are one of these, I recommend that you weigh yourself daily and average it weekly. Here's the way weight loss works:

Body weight = starting weight
plus weight of food
minus body waste products
minus caloric energy expended.

Depending on when you weigh and how much liquid you drank with your meal, your weight can fluctuate several times a day. One 12-ounce can of diet soda will show up on your scale as a pound (scales round up not down) of weight gain for several hours after you drink it. Using a weekly average as shown below will give you a more accurate weight-loss record.

Date	Wt.
Mon.	xx
Tues.	xx
Wed.	xx
Thurs.	xx
Fri.	xx
Sat.	xx
Sun.	xx
Total	**xx**

Divide the total by 7 to get your average weight for that week. If you recorded your weight for only 5 days then divide the total by 5. I keep my weight recorded on a wall calendar near the scale. Incidentally, changing scales can interfere with the accuracy of your records; scales can vary anywhere from 5 to 10 pounds. I suspect that makers of bathroom scales figured out that if their scales were really accurate, nobody would buy them. My wife and I both have participated in and watched other people trying bathroom scales in the aisles of a supermarket or hardware store. Of course, overweight shoppers always decide that the one registering the lowest number is the most "accurate." And since skinny folks don't buy scales, the financial incentive to manufacture a truly accurate, non-adjustable bathroom scale is very slim—unlike the average consumer who takes one home. I use a medical scale because I tend to favor accuracy, but you really need not invest in one. The important point here is the direction in which the scale moves over a period of weeks. If it's going down, The Engineer's Diet is working.

No-Lunch Tidbits

Cholesterol

Cholesterol is produced for the most part by your body, although some does come from what you eat. What determines its level is your overall physical condition—weight included. Think of your body's cholesterol-regulating mechanism as if it were a sort of thermostat. If you eat less cholesterol, your body will produce what it needs to maintain the thermostat setting. So limiting the amount of cholesterol you consume, in and of itself, will definitely not lower the amount of cholesterol in your body. What lowers that level is proper weight loss combined with proper exercise. Remember, no matter what you read or hear, eliminating eggs when you are 40 pounds overweight won't cut it!

Salt

Continue to salt your food to taste. Cutting salt to lose weight is senseless and can be even dangerous. The weight you lose is not fat but liquid, and dehydration can result, which is never good. Salt-free diets should be recommended only for people with serious medical conditions requiring the lowering of fluid pressure. They are not applicable to people in reasonable health.

Blood Pressure

Your skin is like a bag. The more you put in it, the higher the pressure. Lowering your weight lowers your blood pressure—so quit overeating. Stress also raises your blood pressure by creating muscular pressure against your blood-flow network. In other words, tense muscles cause resistance to smooth blood flow. The best way to get the tension out of your muscles is a good exercise program that relaxes you; 30 minutes to an hour daily will do it easily.

If you are overweight and out of shape, it is *not* abnormal for you to have high blood pressure. But drugs used as an alternative to exercise don't address the real problem. Lose weight and get some exercise.

Remember, acknowledge the fire! Don't unhook the smoke detector because you don't like the message.

Socializing without Eating

Many of us overweight types are actively involved in running the businesses, both of America and the world. Good relations with our brethren workers demand social and professional interaction on an almost daily level, and certainly lunchtime plays a significant role at the center of intra- and inter-business relationships. A question you are already asking yourself is: How am I going to function if I can't go to lunch with my clients or associates? Actually the problem is not nearly as complicated as you think. No one is suggesting you avoid restaurants that serve lunch. The temptation to eat will not be triggered unless you start eating.

As part of corporate upper management I am expected to go to lunch daily. I usually order the best coffee, tea, or hot chocolate in the house—often with the cream or a melted marshmallow. I'm not worried about calories because I'm passing up the salad, soup, sandwich, or lunchtime buffet and I have a few calories leeway to enjoy a designer beverage. I do not, however, consume alcoholic beverages; not only do they send negative messages in the current business climate, they also mess with my metabolic and energy balance for the day. (See the section on alcohol consumption.) Also I do not nibble on even a cracker because I don't want to launch my digestive system into an unnecessary and wearing process.

Peoples' reaction to your not eating lunch will range from astonishment to curiosity. If you get into discussion, it should be short. What's there to talk about? I tell folks simply that I'm not interested in dieting or in newfangled theories about the body, and that I eat pretty much the way I think my ancestors ate for thousands of years. Following my simple program, I can allow myself to eat what and however much I want at my morning and evening meals, and I don't need to bother with low-fat or no-fat/no-fun foods. Most of all, I no longer worry about being overweight or having poor nutrition. Someone occasionally interjects something about health foods and special "supplements" the body needs. I answer that it's probably better to trust your body's natural cravings than the magazine rack. And I freely admit that a yen for refined sugar and alcohol is an exception to the rule when making food choices, because these are not natural foods that the body is designed to use regularly.

If you still feel awkward being the one non-eater at lunch, rest assured you will have the upper hand by the meal's end. About the time you're ready to jump up and go back into action for the afternoon, you'll notice your companions' sleepy, guilty eyes staring apprehensively at the door. They don't want to go right back to work; they need to take a nap! Who knows? The next time you lunch with the group, half of them may order only a small salad; and by the third time, several of them could be heading off to the coffee urn with you! The no-lunch lifestyle seems to catch on so quickly that within a few weeks after I began talking about it, several friends and some associates had adopted it. As you'll see by their stories, the results thus far have been unanimous. Those who have given up lunchtime eating have experienced weight loss, improvement in blood pressure, lowered anxiety, and less midday tiredness.

Now you might ask: How about when I've been invited to someone's home for lunch and they've gone to a lot of trouble? I'll tell you what I do; I eat lunch and enjoy it. I don't suffer anxiety or guilt because it doesn't happen very often. Okay, you may argue that your daily life is different, and you're in this obligatory social situation more often than I am. Just remember this: If you're justifying "having to eat lunch" even once a week on a regular basis, then I'd say you're full of it! Lunch, I mean. In the event you are doing daytime eating, and you are justifying it in terms of business or whatever, you might as well go back to the shopping mall to look for more diet books. And while you're there, buy some bigger clothes too, because eventually you'll need them.

Living Leaner Longer

ANY RESPONSIBLE EATING PROGRAM has to be combined with exercise to achieve significant or long-term weight loss, right? Wrong! If you don't eat lunch or snack, eventually you will reach your goal weight—with or without exercise. Here is the part that doctors, nutritionists, and fitness experts don't stress. They don't really even explain it. The fact is, the number of calories you burn exercising is far smaller than the caloric content of the additional food you'll eat because of your increased appetite from exercising. When you start exercising, as you may already know, you need additional protein and carbohydrates to repair muscle tissue and additional overall energy to run your system. It turns out to be harder for most people to maintain diets based on portion control or counting calories when they're exercising regularly.

There are real reasons to exercise your body—just as there are good reasons to not leave your car sitting for months at a time—but weight loss is not one of them. Just as your car needs frequent tuning and regular running of the engine to avoid rust, keep hoses flexible, lubricate parts, and recharge the battery, you should exercise your body regularly to retain bone mass, increase and maintain flexibility, build muscular strength, stimulate circulation, and stimulate and strengthen your heart. Exercise is great also for focusing the mind on something besides worrying about how to lose weight! I'll quickly mention four key areas of the body to exercise and point out a few specifics; but I won't dwell too long on exercises. After all, this is not the book to end all books on how to become Mr. or Ms. America, make it to the Olympics, or prepare yourself for a new relationship (although you may be reading this book in order to lose weight because you're planning to accomplish any or all three of the above!). But remember, this book is really just a simple plan for weight loss.

1. Bone mass (extremely important for those over 40) can be maintained very simply with free-weight lifting: for example, three sets of repetition using relatively light weights, three times a week. I've included my own exercise program at the end of this chapter, but those of you who are weight connoisseurs or experts know where to find programs you like. My intent here is simply to tell you how to get a general workout.

2. Flexibility is best achieved when your muscles are warm. So get the blood flowing to your muscles before you stretch. You'll find that ballet dancers, martial arts experts, and aerobics instructors all base their workouts on this maxim. I don't lean toward regular aerobics classes myself; they are pretty new and look to me more like Hop-and-Puff-ercise. Rather I recommend martial arts (something like Tae Kwan Do) or ballet; these movement forms are at least several hundred years old. They exercise the whole body and also include an aesthetic component. You can pick up some of these techniques for practice at home; or, better yet, take two non-lunches a week and attend a dance or self-defense class. You'll quickly notice the difference in your appearance as well as how you feel. What we regard as looking "old" has more to do with flexibility than you might realize; losing flexibility in the backs of the legs, and the ability to turn the head, will give you an aged appearance, even if your actual features are quite youthful. And here's another tip: The "oriental" tradition of bowing rather than shaking hands stretches the large muscle sets on the backs of your legs many times a day. It also prevents the spread of germs or disease through handshaking!

3. Keeping your arteries or "hoses" open, stretched, and in good shape involves running your blood through them at varying speeds and pressures. Good artery health is critical in avoiding the disabling conditions that call for such elaborate procedures as bypass surgery. You can maintain healthy arteries by the type of weightlifting that includes repetitious muscle-pumping—an aerobic workout. Aerobic exercise elevates your respiration and heartbeat enough to get more oxygen into your blood and should be part of any long-term exercise program. But there's room for question about how often to do aerobics and for how long. Many people attempting to exercise in order to lose weight have gotten themselves into exhausting programs of long-distance exercise, which ultimately they can't keep up. Programs like running or cycling, in and of themselves, don't result in sustained weight loss and can cause other kinds of problems—pulled muscles, joint deterioration, or painful sprains.

4. If you accept the fact that lunch is the cause of your weight problem, then you're also ready to consider the engineering rationale for exercising properly. An engineer's view of your heart is somewhat different from that of a medical guru. Basically, your heart is an electronic pump with a spring diaphragm. That means that "coming off the shelf" it has a limited number of potential shift cycles. Its potential also has to do with who the manufacturer is—in this case, your parents. One thing is certain, though: while you cannot *increase* its potential, you can *decrease* it in a number of ways.

Like all the other organs in your body, the heart has a maximum potential. When exercise experts talk about strengthening the heart muscle they're not talking about increasing its potential. Remember, there's a maximum life span that was determined at your birth. Literally, from the day you were born you have been running down. You can perform lower than your potential but never higher. If I go out and buy a pump, I can maintain it and get the maximum life out of it. Or I can take poor care of it and get only a half-life. But I can never get more than the life it has. Your heart pump is something like the pump in your car. If you don't use it at all, it freezes up and ruins the engine. If you use it too much, you'll wear it out early. So you want to use it just enough to keep it in good shape. Unless you want to be in the Olympics, in which case there's a trade-off—high performance vs. maximum longevity. I'm not arguing either one, but we should try to understand clearly how the system works.

Metabolism and Exercise

When you're young, as we pointed out earlier, you continually burn energy, but your body is equivalent to a small, economy car. High-performance cars, if they're small, get good mileage. When you get to be a big, old truck, you probably don't want to race the engine as much because now you're a gas hog. Understanding that you're either a big car or a little one makes all the difference. In adults, pushing up the metabolism through exercise, except for specific, short periods, is not a plus.

As I said earlier, metabolism is the rate at which you consume fuel. You have a specific metabolic rate even when you're not exercising and instead maybe sitting, reading, or sleeping. The motor is not off; if it were, you'd be dead, but the engine is idling. In fact, the resting rate is your regular idle. You're not doing heavy work, but your engine still has to run—your air-conditioning is on; your eyes are processing information, and so on. It's mostly chemical and electrical activity on the inside—think of it as your computer on screen saver.

When you start moving around, energy requirements for any exercise or work are increased significantly. That means you're going to need more fuel. You can run up your metabolism by doing exercise but it's highly unlikely that you can exercise long enough to burn off the extra fuel you'll need to consume. This is the part that you don't hear much about in the ads for health clubs and workout studios. Start taking a one-hour aerobic class, and watch what happens. Your metabolism speeds up, and you eat more. Even if you increase calories by only 100 per day—an unrealistically low number—it's enough to cancel any weight-loss benefit. Let's assume your intake averaged at least 2,000 calories per day. Add 100 more. That's 14,700 calories a week. To lose a pound, you need to "burn" off 3,500 per week or 500 a day, every day. This is next to impossible for a non-athlete to do in one hour on a consistent basis, and, if you increase the exercise, you'll more than likely increase caloric consumption as well. At some point, you may trade off feeling good from a reasonable workout for running your entire system down just to metabolize the extra food! That's aside from the muscle tissue that you're tearing down through exercising, which may require that you consume certain foods you may not have been eating before. Then there's the question of whether you want to raise your heartbeat that high and for that long.

Don't "Pump It Up"

Assuming most of us are not gold medalists, we're after health and longevity. Simply stated, the longevity of the heart rests on finding the optimum beats-per-shift cycle. It would be a "no-brainer" if the ideal shift-cycle required that the beat never sped up. Then we could sit around with the slowest possible heart rate and expect the longest life. But reality is slightly more complex. Without some exercise to speed up the rate at which blood pumps through the heart, the system can become somewhat clogged and inefficient. In numbers that could mean, for example, that the heart works harder, beating 72 to 80 times per minute, than it would have to if it were in shape, beating at 54 to 66 beats per minute. So, although the goal theoretically is to keep your heart rate perpetually low in order to maximize its limited potential cycles, the only way to keep your heart in shape is to expend some beats by operating at a higher rate for a brief period daily, this improved performance enabling you to operate at a lower rate the rest of the day. This numerical reality is the key to designing your aerobic workout. Exercise long enough to lower your heart rate for the remaining hours of the day but not so long that you use up the beats at a faster rate. Remember, used beats cannot be "recovered" in either exercise or non-exercise hours. Here it is in mathematical terms:

No Exercise Day Heart Rates	
Heart rate per minute	80
Heart rate per hour	4,800
Total heart beats per day	**115,200**
Exercise Day Heart Rates (1 hour of aerobic exercise)	
Heart beats for 1-hour exercise session	8,400 (at 140 beats per minute)
Heart beats for 23 hours at rest	82,800 (at 60 beats per minute)
Total heart beats per day	**91,200**
(2 hours of aerobic exercise)	
Heart beats for 2-hour exercise session	16,800 (at 140 beats per minute)
Heart rate for 22 hours at rest	79,200 (at 60 beats per minute)
Total heart beats per day	**96,000**

Looking at this math from the perspective of the life of your "pump," we see plainly that exercising for 1 hour a day is better either than none or 2 hours per day. Of course, this is if you're not engaged in the profession or avocation of competitive fitness or sports. For non-athletes, the overall data on exercise have consistently pointed to regular, moderate exercise as the best guarantor of long-term health and fitness. Almost everyone agrees that walking is one of the best aerobic exercises and that two to three times a week is sufficient. Somewhere around 45 minutes to an hour at a time, for example, on Monday, Wednesday, and Friday.

Including a moderate weight program on alternate days would be an excellent plan. For those who want to do a little more and who wish to know more about aerobics specifically, I recommend looking at Ken Cooper's first book, *The New Aerobics*. I have found him to be a real guru in the field of aerobic conditioning.

My Own Weight-Lifting Program (Tuesday, Thursday, and Saturday)

	Sets	Reps	Wt.
Wrist Curls	3	10	8–12
Reverse Curls	3	10	8–10
Behind Head Press	3	10	10–30
Arm Raises	3	10	8–20
Leg Lifts	3	10	0
Presses Single Bar	3	10	50–100
Presses Dumb Bells	3	10	20–40
Curls Single Bar	3	10	40–90
Curls Dumb Bells	3	10	20–30
Toe Raises	3	25	0
Lunges	3	10	0
Crunches	3	10	0

For the remainder of my exercise, on Monday, Wednesday, and Friday, I do about 3 miles in a combination walk-run. On Sunday, I usually rest.

Other Consumables

Drinking as a Mechanical Problem

For years, every time I picked up a new diet book, I flipped to the index to search for alcohol and cigarettes. Maybe, on a subconscious level, I was looking to see if either one would be allowed. But the plain truth is that both are intimately related to effective weight management. I've pretty much decided that any diet or health program that doesn't address drinking alcohol and smoking nicotine is not going to work for very long for those people who do either or both.

Alcohol is one of the most interesting, problematic, and paradoxical of all consumables and contributes a great deal to many people's weight problems. We've all heard about the "empty" calories in alcohol. That's a bit misleading, because they're really just the opposite. An alcohol calorie is the "fullest" of all. It is literally high-octane fuel. You can run your car on it! But it can also create a potential problem for your body. You get instant energy when alcohol enters your system, but because it's actually sugar in its most refined form, the energy effects are also the shortest. And the paradox is that even though alcohol will put weight on us, abstention from drinking will not necessarily take it off—in and of itself, that is. Remember, you need to be as far under your daily maintenance level in calories to lose weight as you have gone over it while gaining. And in that sense, it doesn't really make any difference what kind of calories we're talking about. What will differ, depending on whether you choose fat, protein, carbs, or sugar, will be the length and type of energy effects.

Perhaps you're thinking that if alcohol is an important part of your life, you should be able to work out a plan that skips lunch but doesn't exclude a daily cocktail or two—either with or after dinner. Well, you could do that, because in fact, whatever you're now doing (if it includes three meals a day) will be improved if you skip the meal at noon. But from an engineering point of

view, although alcohol will temporarily "gun your engine," it will ultimately "screw up the timing." Of course, it's true that one or two beers—because of the greater proportion of carbohydrate compared to hard liquor—added to a stomach "tank" filled mostly with lower-octane fuel—a varied combination of nutrients—won't throw you completely off of a good program. But, let's be straightforward about this; for more than 10 percent of Americans this is not what's happening.

I hesitate to use the term "alcoholic" here for a number of reasons—not the least among them the current social tumult surrounding the merits or problems associated with drinking. And in this book, I'm not going to argue either side, so I'll just refer to drinkers instead and give it the broadest possible definition. In other words, for my purposes, you are a drinker if you take even one drink more than a toast at, say, Thanksgiving, or your daughter's wedding.

People drink for multiple reasons. Alcohol provides that shot of quick energy, as well as being a relaxer and a painkiller. However, as I indicated earlier, more than 30,000,000 Americans have developed a tolerance to alcohol that nullifies at least some of its sedative qualities—and you know who you are. You handle alcohol better, drive everybody home, and still hang in there. And perhaps you don't think of yourself as a "problem drinker." But being blessed with a sedative-resistant body, the fact is that you need to consume large quantities of alcohol to get the effect you desire. The result is your entire engine performance goes berserk because it's difficult to maintain your optimum fuel balance.

If you are one of those with a high tolerance for alcohol, and you want to start The Engineer's Diet, I have good news and bad news. The bad news is you need either to abstain totally or limit your drinking to a couple of events a year. (I might add that an event isn't from Thanksgiving *through* the New Year; it's more like the number of times that you come out of the house and find your car's been smashed flat by a meteorite.) The good news is that by eating only twice a day, you cut down the temptation to consume sugar in any form—including alcohol. And if willpower is a problem for you, don't be embarrassed. Alcohol is not a natural food so your body can't be expected to react to it in a natural way. For help with your willpower, there are various programs—some of which are spiritual like AA. But you may not like or need them. You should also know that in contrast to what is reported by AA members and many drug and alcohol counselors, there are apparently millions of Americans who perhaps drank excessively at one time or another and now either don't drink at all or drink only once in a while. I happen to be one of those who finds it easier to abstain entirely when managing my weight properly. I also got a lot out of a book I read called *Under the Influence*, by James R. Milam.

To Smoke or Not—That Is the Question

Smoking is a tough one. Nicotine is perhaps the most addictive substance on the planet. Look at how many alcoholics and drug addicts have given up terrible addictions—especially things like heroin and crack—and continue to smoke cigarettes! Add to that the fact that most people who quit smoking start gaining weight and take up smoking again because it seems like the lesser of two evils. Surprisingly, it turns out that they could be making the right decision—if those were the only two choices. I remember reading research studies in the early '60s that showed obesity to be a bigger killer than smoking. Even as most people's bodies led them to intuit that smoking was less dangerous than overeating, the medical profession, in its infinite wisdom, went after smoking with a vengeance. This is the more surprising when you consider that studies have failed to show lung cancer is significantly higher in smokers than in non-smokers. As of now, the American Lung Association and other interests are blaming secondhand smoke rather than reevaluating the entire premise. If they add secondhand smoke to the list, they can make a good case for smoking as the primary cause of cancer —even if the data are not overwhelming.

It is of interest to note that longevity is not particularly lower in European countries, where people continue to smoke—in fact they may be outliving us, if I'm not mistaken. It does seem that obesity can be more of a strain on your engine than air quality is. It happens that I no longer smoke, nor do I recommend it. A good engine should run on good fuel and draw clean air into the carburetor. But I have to be honest here—from a mechanical point of view, given a choice between adding a significant load to the vehicle or traveling light with less-than-clean air—I would bet on the longevity of the lighter vehicle every time. Fortunately you don't have to choose. You can travel light *and* have smokeless air if you just don't smoke and eat so much. Again the best advice comes down to eating right—no lunch, no daytime snacks. And if you do manage to get off cigarettes, try to stay off. Every serious-thinking grownup I've talked to recently agrees that good weight management and a life relatively free of addictive substances are the way to go.

Chemical Imbalances: Are They Real?

Are there particular foods that help to solve emotional problems for the 30 to 40 percent of the population who now believe themselves to have a chemical imbalance in the brain? I want to address this subject because I believe there is a dietary connection here, although nothing like what's being presented most of the time. I'm also tired of the confusion created by an overload of definitely questionable, and potentially harmful, misinformation that we are being fed about how our our brains work. Even more frightening, however, are the drugs we're being fed—Prozac, Zoloft, Paxil, Xanax, Lithium, and a host of other antidepressants and mood regulators.

Twenty-five years ago, the experts told us that depression, phobias, and high-level fear or anxiety were caused by childhood experiences. In most cases, the treatment of choice was to relax the person and, using various types of psychological therapy and verbal analysis, take the individual through a process of reliving or re-experiencing—a sort of rewinding of the tape. This allowed you to understand childhood experiences differently and to gain new coping skills to deal with present problems. Then, as the medical world probed the human brain and claimed to understand its workings, there began a shift toward replacing traditional psychotherapy with drug therapies. The stance has been that certain chemicals are "needed" to affect the actual brain and to establish a foundation for the treatment process. The accompanying talk was, and still is, all about chemical imbalances in the brain.

At the risk of sounding cynical, I must state that I suspect the "success" of drug therapies has somewhat less to do with solving the patient's problem than with furthering a successful joint venture between therapists and pharmaceutical manufacturers. Nor do I find reason to believe most of the latest psychological theories or any of the new brain theories. First off, they're operating on a main assumption that the human body isn't properly designed for living. They dare to claim that, even after millions of years of evolution, your body and the world it lives in are at odds—so much so that you must be "chemically altered" in order to survive! Are they serious?

The argument is that the pressures of modern life have introduced new stress factors that you can't cope with on your own. Well, I don't buy it. Contemporary literature has acquainted us with the postmodern, fractured, rat-racing, transnational planet we've created and now can't manage. How can we step back from all this hype? I did some of my growing up in a rural, relatively primitive, desert community and experienced some major stress: How about fear of rattlesnake bites or angry farm dogs that never had a rabies shot? And the grownups around me were dealing with plenty of stress as well—drought, famine, and a limited selection of mates, to mention just a few. Who's to say that these pressures are any less intense than the ones we face today and that are driving so many people to seek drug therapy? Actually we face similar challenges—contracting a deadly disease, making enough money, competing for sexual partners. What's so special about now?

On top of that, therapists now are telling us that it was "breakthroughs" in brain science that led to the miraculous discovery of chemical imbalances. This is the same brand of professional who was so convinced that your problem was caused by emotional trauma during childhood. How did these experts suddenly become converted? Did they go back to school for intensive study on new data about the brain? No! Are they prescribing drugs now on the basis of something they saw on the Discovery Channel or perhaps read in a journal article? I certainly hope not, but I'm suspicious.

Finally, let me focus just briefly on the premise that a chemical imbalance in the brain is, in fact, a problem.

That is, assuming that there is such a thing (which can't be proved) and further assuming that the doctors understand how it functions, which you can be pretty sure they don't. At the outset, a chemical imbalance of the brain sounds serious—like maybe those odd thoughts you're having are about to take you over the edge of a cliff unless you somehow get chemically balanced. The medical gurus can make it sound like your brain has just suddenly gone haywire and has to be fixed. But notice what they don't say. Somehow they have forgotten to mention that your brain can't even function in a real state of equilibrium. In other words, without a chemical "imbalance" of some sort, you wouldn't even be able to think. What the doctors are suggesting by prescribing drugs to "balance" your brain may be roughly equivalent to your auto mechanic wanting to remove the battery so you no longer notice the electrical problems your car's been having.

Think in terms of voltage. If you're one of those rare folks who stays busy in a satisfying career—exercising, eating right, not on drugs or over-drinking—and your life still seems in turmoil, I suppose you might want to consider "turning down" the voltage. But the majority of people are not in this situation—aren't eating right or exercising—and they're probably not engaged in fulfilling work, at least they don't act like it. For most of us, it appears that brain imbalances are not only appropriate but inevitable, and exist almost by definition. As I said earlier, you don't take the battery out of your smoke detector because you don't like the idea of a fire where there's smoke. Since the history of human accomplishments almost reads like a catalogue of brain "imbalance" I suggest that you don't tamper with brain function when you become tired or depressed and need to re-assess or modify some of your activities. Maybe the same brain that generated your accomplishments can generate your healing. It's worth a try.

So back to the original question: What kind of foods do you eat to arrest the chemical imbalances in your brain? Nothing in particular, because as I see it, you probably don't have any. In other words, probability is that the brain you were born with is fine. Common sense should tell you that without the influence of drugs, alcohol, or excess sugar, a very small portion of the population has congenital brain dysfunction—perhaps about the same statistical proportion of folks that are missing various limbs from birth.

I am not a physician or a psychiatrist, nor am I trained in either of those disciplines. My opinions on this subject are based on engineering logic. For instance, take hyperactivity in kids. Instead of feeding them drugs to slow them down, why not get them moving? Moving in constructive and creative ways. You can trust the human body to react well to common sense. And for those of you who have already "unhooked" your warning lights to get through a difficult period, or because you believe that chemical imbalances are unnatural, I urge you to work with your therapist to go down to minimum drug therapy as soon as possible. Remember also that most of the drugs on the market are too new for anyone to have seen the long-term consequences.

I'm not arguing in favor of suffering, but I do suspect that 99 percent of Americans should probably not be diagnosed as having a chemical imbalance. The primary emotional conditions attributed to unbalance, such as depression and anxiety, may in fact be both "normal" and even essential to maintain our overall health. While it's true these conditions have long been regarded as signals of something, the experts are confused about what the signals are saying. Anxiety, for example, might be looked at in terms of what it's trying to tell you about something in the future: Some plan you are part of—something you think or know is going to happen, or that you have to undertake—and you're nervous, scared; that is, anxious. What's so abnormal about that? The anxiety will leave if you chose one of two options:

1. You cancel plans or figure out changes that will help you to avoid doing what scares you.
2. Go through with it and realize that you have survived.

Either way, the anxiety has served a valuable alerting function and should be short-lived. From an engineer's perspective, there doesn't seem to be anything so complicated about it. Why medicate? It's similar with depression, but even more interesting. If I suggested that depression in the human had a corresponding mechanical function in the auto engine, what would it be? If you guessed the clutch, you were right. Gear changes can't happen without disengaging the clutch. Depression is a signal that you have to stop and look at something present in your life that is making you unhappy—your job, your relationship, the weather where you live, or your physical condition.

A lot of it goes back to eating right. In fact, I would bet that if therapists were willing to consider it a valid basis for "depression," they would find that the majority of depressed patients are unhappy with how they look and feel. In other words, the bottom line is that most likely people aren't overeating because they're depressed; they're depressed because they're overeating. I mean what I'm saying. The logical solution in such cases would not be taking medication; it would be changing the circumstances behind the unhappiness.

It gets back to trusting your body's instincts. Pharmaceutical companies, and the diet industry as well, stand to gain a great deal by convincing you that neither your body nor your mind can be trusted to properly manage longevity, health, or the enjoyment of your life. If they convince you that you can't do it alone, they maintain a trillion-dollar industry. But, whether you believe in natural selection or in spiritual theories of how humankind came to be, in either case, there's no rational justification for human beings being dependent on pharmaceutical enhancements or dietary substitutes. Those who would tell you otherwise are likely engaged either in consuming these types of substances or in profiting from them. And even if there are exceptions where these treatments are necessary, the cost would usually keep them out of reach for middle-class America, and special resolutions of the problem would have to be found.

Are You Grounded?

To speculate a bit further on how little scientists really know about the human brain—here's a question. Experts, including medical ones, would have you believe that they understand fully how neurons within the brain function in conjunction with emitters and transmitters to process information. Sounds pretty intimidating—especially when they borrow computer, solid-state language to describe human thought. But an engineer like me has an issue with these guys who supposedly understand complete brain cell function, but can't yet make a single living cell.

My question is this: How come, with all the talk

about microvolts and electrochemical processes in the brain, nobody ever talks about grounding? The human body is a reasonably good conductor—which incidentally was continuously grounded until people started wearing shoes. Isn't it possible that the "unnatural" amount of static electricity that builds up in the modern human body can create problems for the brain? We intuitively understand the need to ground the stereo, which is made up of far less highly sensitive components. Why is no one concerned about grounding the human body? From an engineering point of view, it makes perfect sense that so many monks and gurus wear no shoes. I find it reduces tension considerably to take off my shoes and walk around outside. Maybe there's more than an idyllic memory to the barefoot days of our youth.

My advice: If you're experiencing a lot of tension, anxiety, or depression—a kind of "electrical noise" in your life—try grounding yourself before you head off to the doctor. It only takes a minute to take off your shoes and socks and find a place where there's no carpet. Of course if you're trapped in an upper-story New York apartment with wooden floors, you'll have to prop your feet up on some water pipes that run down to the basement. Who knows? Maybe after a little grounding you'll straighten your thinking enough to leave your city life for a while and take a spell in the country where you really can get grounded by putting your bare feet in the mud! The mud along riverbanks, when you're fishing in the shade, is ideal for grounding the static buildup in your brain. Even if it doesn't solve all your problems it'll give you a break from crowds and concrete. And it makes good electrical-engineering sense.

The Engineer's Checklist

WE'VE ALL HAD FRIENDS AND RELATIVES who are caught up in the weight-loss yo-yo. They try all kinds of new diets, share them with us, we try them and pass them on to others—though few of us ever achieve success. I suspect the popularity of the latest diets has less to do with the soundness of their theories than with a desperate need for faith on the part of most overweight people.

I'll bet you've tried some bizarre stuff; I sure have. Diets like citrus all day, followed by all green vegetables, then all bananas, then a day of nothing but steak. Sounds crazy doesn't it? But actually such a diet would be merely a simulation of the natural diet of our ancestors—the guy is walking back to the cave from picking fruit and gets chased through the tall grass and up a banana tree by a large animal, which he subsequently kills and eats. My point here is you can prove just about any diet has historical roots, because in fact, if it can be swallowed, then human beings have probably eaten it.

Comparative Summaries

At this point, I think it would be helpful for us to compare The Engineer's Diet, a sensible, two-meal-a-day, trust-your-instincts, eating program, with some of the popular diet solutions on the bookstore shelves. Incidentally, most of the current bookstore chains provide coffee areas, which will allow you to maintain your Engineer's Diet as you browse through the more complex books on the subject. I call these books the maximum demand–minimum results solutions. The information below is my take on a few of the diets that seem to represent prevailing philosophies of modern eating. Each diet book is graded on the basis of the following six attributes:

1. Rationality—Nutritional logic on which the diet theory is based.
2. Desirability—Appeal of the foods recommended on the diet.
3. Sociability—Desirability of sharing eating experience with others.
4. Availability—Ease of attaining required foods (i.e., transportation or travel).
5. Convenience—Time and trouble required to prepare meals.
6. Cost—Additional cost of specialty foods or services.

The Zone by Barry Sears

1. Rationality. According to the nutritional logic behind this diet, we are in The Zone when we are in optimum health. So far so good; I like that and I get

kind of "zony" thinking about perfect health as being able to throw 90-mile-an-hour fastballs, for example. But by chapter 2, the book's logic moves from zony to phony. The premise: Because high-carbohydrate eating didn't solve America's weight-loss problem, we need to eat more protein and fat. This argument has a peculiarly familiar and somewhat tiresome ring to some of us older weight-loss failures—we see it as the rebirth of the Scarsdale diet from years back. (It is my personal opinion, in fact, that Mrs. Scarsdale might have come up with a good insanity defense if she had pleaded the oppressiveness of her husband's all-protein diet.) At any rate, a diet such as The Zone, which highly restricts carbohydrates in exchange for exacting ratios of fats and proteins at every meal, is clearly not based on any seasonal availability of foods that our ancestors experienced. But the medical jargon is impressive and possibly of some merit—that is, if anyone can maintain the required lifestyle—so let's continue looking.

2. Desirability. I found my mouth watering as I read through the various gourmet suggestions, so this diet certainly recommends good foods!

3. Sociability. I'm afraid unless you want to spend the rest of your life talking about protein ratios and carbohydrate blocks, this is not your diet. Just the busy activity of organizing your plate, when you are a guest at someone's table, would at a minimum distract the other guests, if not insult your host.

4. Availability. The diet food can be acquired at some of the finer restaurants, but it does not allow for a simple breakfast of cereal—unless you can find a handy morsel of smoked salmon in your cupboard to offset the carbohydrates!

The Engineer's Checklist

	The Zone	Eat to Succeed	Get Fit	Dr. Atkins	New Pritikin	The Engineer's Diet
Rationality	absurd	absurd	absurd	doubtful	doubtful	outstanding
Desirability	out-standing	okay	absurd	okay	okay	outstanding
Sociability	absurd	absurd	doubtful	doubtful	doubtful	outstanding
Availability	doubtful	doubtful	doubtful	pretty good	pretty good	outstanding
Convenience	doubtful	doubtful	absurd	okay	okay	outstanding
Cost	absurd	absurd	absurd	doubtful	pretty good	outstanding

5. Convenience. This is not an issue if you share Dr. Sears's credentials—and income—and can hire a full-time cook. The meals are complicated and time-consuming for regular folks.

6. Cost. This is another area where The Zone will make you groan. Beef Szechwan or "Eggless" Egg-Salad Plate are just *not* what pop into my mind—or my oven—when it is my turn to make spaghetti for the kids. The costs of making meals on this complex, restrictive diet are far beyond the means of most Americans who still love to eat affordable foods like potatoes and bread.

I have to conclude that The Zone is a finicky diet for the wealthy—demanding in both time and money. Also unnecessarily restrictive, it is based on questionable scientific conclusions and is difficult to maintain. Without an analytical balance (a chemist's scale) and some other rather specialized equipment found only in laboratories, I am not sure how you would know when you were actually on it. Its greatest strength is that when you fail to lose the weight, you share this experience with 90-plus percent of the population who can't even understand the diet enough to try it.

Eat to Succeed by Robert Haas

This publication should not be read as a diet plan for ordinary healthy people. It's written for athletes and for particular individuals who have not managed to gain weight on three meals a day. There are only about ten pages in chapter 6 that actually address weight loss: his "maximum weight-loss program." The program is based on especially formulated, liquid meal replacers, carbohydrates, and fat-burning, high-energy nutrients such as 1-carnitine and coenzyme Q_{10}.

1. Rationality. The nutritional logic behind the book is that with 200 pages of recipes and a better understanding of blood chemistry, you will be more successful in your career. As an engineer, I question whether constant preoccupation with food and eating—a compulsive disorder by some standards—would give me enough clarity of mind to make up for the thinking time I lose fixing all those complex meals! Haas's diet is highly restrictive and, like The Zone, requires the constant balancing of foods and your continuing to distrust your body's natural preferences.

2. Desirability. I suppose it would taste okay—with the help of a cooking staff, and it does allow for more carbohydrates than The Zone. But the liquid supplements are questionable and highly unnatural to ordinary human appetite.

3. Sociability. I would think this diet could be hidden to some extent from your friends—as long as they didn't see all those liquid supplements in your car or lunch box. Personally, I am not interested in lugging cans of liquid supplement to the company picnic. As a discussion topic, this diet would be so boring and restrictive that, in fact, I'm even having trouble summarizing it in this book!

4. Availability. It requires considerable extra shopping and complex meal planning.

5. Convenience. You would need to eat three times a day at a minimum. And when eating out, you might need to enroll the chef in investigative research and planning, in order to adjust the proper ratios between fat, protein, and carbohydrates.

6. Cost. The diet is full of gourmet recipes, making it highly unaffordable for ordinary Americans who have both time and budget limitations.

Overall, the Eat to Succeed diet seems to be a sort of apology for the author's earlier, high-carbohydrate *Eat to Live* book. Haas implies that rapidly changing knowledge about nutrition is the reason the second book had to follow the first. This book may have some application for world-class athletes who do not have weight problems and can dedicate all their spare time to food theories and adjustments for minor improvements. But the book has no relevance to us ordinary, overweight folks.

Get Fit by Larry North

1. Rationality. The Larry North program takes compulsiveness to a new level. As president of the Texas Council on Compulsive Gambling and the son of one of the founding members of Overeaters Anonymous, he has built an entire lifestyle around eating five or six times a day. I know someone who lost 20 pounds on the Larry North diet, but got tired of carrying around and eating the six daily meals and gained the pounds all back. The logic behind this diet is to raise your metabolism and get your weight down through eating constant low-fat or no-fat/no-fun foods, combined with heavy exercise. Besides being, in my opinion, just plain idiotic and unnatural in terms of your life span, this program goes against your body's built-in conservation system.

2. Desirability. The Larry North diet scores very low here. You are discouraged from many foods, including

fruit juices, cereals, white flour, pizza, and so on. If you're at all like me, this diet seems designed for masochistic compulsives.

3. Sociability. Anyone who stays on this diet for very long won't have to worry about socializing. Eating six high-fiber, low-fat meals a day, in combination with dried turkey chunks, does not put you in a very sociable mood.

4. Availability. Not much except in Dallas—North's home town—where several restaurants are so sympathetic to people who can't eat regular food that they have put a North Plate on the menu.

5. Convenience. Six meals a day? No Comment!

6. Cost. The additional cost of eating six times a day, not to mention the cost of supplying separate meals for your kids, could be overwhelming. You won't get your kids on the North diet because they are far too sensible to give up ordinary foods—like bread.

The North diet has only one thing going for it; the justification of compulsive behavior. Normally, to constantly recheck your stove, refrigerator, or furnace could be regarded as symptomatic of an obsessive-compulsive disorder. But with the North diet, you can argue that your unusual preoccupation with the stove is merely checking the progress of your six required daily meals. In all seriousness, this diet is tremendously wearing on your entire system and breaks every single rule for preserving and maintaining equipment in good order. Considering that the equipment we're talking about is your body, I would be wary of adopting the North lifestyle.

Dr. Atkins Diet Revolution by Dr. Robert Atkins

1. Rationality. The rationale behind this diet is that humans are basically carnivores except for their consumption of milk and cheese. If this were in fact true, neither Native Americans nor Chinese have ever had a natural diet, since they have no history of using dairy products until very recently. Also, such a supposition is not borne out by our jaw and teeth structures, which were designed for grinding grains (carbohydrates) with our back teeth. Finally, the diet does not permit the human body any access to vitamins through fruits.

2. Desirability. Steaks are fine, but after awhile, meat and cheese get a little boring.

3. Sociability. Not a problem. On slow-burning fuels like protein and fat, you will remain in a bad mood and never feel like going anywhere. Also, the bad breath that results from continuous burning of meat and fat, will protect you from social interaction with most people.

4. Availability. The foods on this diet are usually available. However, you will need to avoid a good portion of most prepared meals, since for most of us, carbohydrates are considered essential elements of the human diet.

5. Convenience. Here, too, you will have to separate the kids' meals from yours. High-protein diets are not only unappealing to most youngsters but have actually been found to be somewhat dangerous to their overall health.

6. Cost. The cost of constantly eating steak and cheese is substantially higher than most people can

afford. Additionally, the cost of dialysis resulting from kidney damage—due to long-term, high-protein diets—needs to be factored into your budget.

Atkins's diet has been shown to produce some early, rapid water loss, and it does simplify eating by reducing it to two food groups, meat and cheese. If you're lucky, your body will resist staying on this diet for too long because it does in fact have severe consequences involving your long-term health.

I suppose one pleasant by-product of the Dr. Atkins Diet is that it allows you to throw "ketostixs" in the toilet. The idea here is that if the sticks turn purple, your body isn't getting the carbohydrates you need to convert your protein and fat to energy—ergo, you are digesting those elements for no purpose. If you can set aside concern about the wear and tear of eating food you can't metabolize in the first place, you may have some fun playing at making pretty purple colors in the toilet bowl. But could I suggest as an alternative that you skip lunch instead and use your lunch money to buy a watercolor set? That way, you can work with the entire spectrum of colors!

***The New Pritikin Program* by Robert Pritikin**

1. Rationality. This "new" Pritikin diet is a portion diet that will keep you fairly preoccupied with your eating habits, but at least it does not attempt to manipulate the protein/fat/carbohydrate ratios. The weight-loss rate seems sensible in that it suggests a lowered calorie intake of 1,000 calories a day through the weight-loss period. However, Pritikin's approach is

based on requiring you to avoid eating full portions. Therefore, it turns out to be scientifically irrational because it demands that humans forgo their natural inclination to eat at each sitting until they are full. For people with healthy appetites and a love of food, this diet seems doomed to fail.

2. Desirability. The diet does allow for balanced eating, so you can enjoy a wide range of your favorites. However, you've got that problem of leaving in the middle of your meal. We talked earlier about that one! Remember Fido?

3. Sociability. You can eat this diet anywhere, but you may find that you're grumpy from never getting enough to eat at one sitting.

4. Availability. Because of diversity, the Pritikin diet does pretty well in this category, but the warnings about salt and cholesterol seem superfluous; you don't have to worry about these anyway, once you lose weight. Also, on this diet you have to do a sort of stop-and-go traffic control when you go out or eat with friends.

5. Convenience. The diet is actually not very inconvenient—primarily because you won't be on it for very long. See, it forces you to control portions and you and I are not inclined to do that.

6. Cost. Additional costs are minimal. But you may develop an aversion to having to pay for full portions when Dr. Pritikin is not allowing you to clean your plate. Of course, you could try convincing restaurants to give you a cut-rate price for your smaller-sized orders.

Basically, The Pritikin diet would be okay, except that it does limit portions for all three meals. In fact, his diet book includes an entire section on guilt and emotion in anticipation of his readers' failure to get up and go away in the middle of their meals. Someone might suggest to Pritikin that his diet, with larger portions and less frequency, would be very manageable and healthy. Maybe by skipping lunch and eating a good full breakfast and dinner.

The Engineer's Diet by Thomas Shaw

1. Rationality. The diet was designed by an engineer. If you have ever known one, or you have read this book, enough said.

2. Desirability. Eat whatever you want twice a day—except watch overconsumption of sugar and alcohol. It doesn't get much better than that!

3. Sociability. On this diet you will actually lose weight, and that will add tremendously to your self-confidence. Also, you'll have the freedom to enjoy food at dinner without measuring portions, calorie counting, or picking your food apart. This means that the subject of diet will never come up as a topic, except at lunch—at which time people will ask questions. You will be in control of the conversation, because you'll be the only one free to talk. The rest of your friends will have a mouthful of food!

4. Availability. You will have absolutely no trouble finding "no lunch" anywhere on the planet. Besides that, you can eat this meal anywhere from the Sahara desert to the Vatican without dropping a crumb! And of course, at breakfast and dinner you are free to eat whatever's available.

DIET BOOKS
EAT TO SUCCEED
ATKINS DIET
?!
The DIET DIET
THE ZONE
DIET
The NEW PRITIKIN
GET FIT!

5. Convenience. You'll save time and energy preparing only two meals a day. Also you will not suffer the embarrassment of new food stains to your clothes during the day.

6. Cost. This is obviously the most cost-effective of all the diets.

* * *

There are numerous diets currently available that recommend limiting either fats, proteins, or carbohydrates. In contrast, limiting yourself to two portions a day allows the greatest flexibility in how much you eat and what you eat—while dramatically reducing your overall caloric intake for the rest of your life. If you look at the checklist, you'll see that The Engineer's Diet is a clear winner in all categories. Criteria such as cost, convenience, availability, are essential to accomplishing long-term results with any program that you start. Diets you can't maintain over the long term provide no significant improvement to your physical health and can even end up being destructive to your mental health. It is my hope that you have found some of my thoughts and experiences useful. Whoever said there's no such thing as a free lunch, it wasn't me.

Personal Stories

Robert H.
Male, age 64
Progress after 15 weeks—weight loss 13 pounds

I remember the one time when I was noticeably overweight, which was back when I was running a company in the 1970s. I was in my early 40s and got up to over 200. I stopped eating lunch for awhile and my weight went right down. I didn't think anything about it. I must have returned to three meals a day fairly soon after that. Didn't worry about weight again and didn't skip lunch again either. But, like most people, I started gaining 1 to 3 pounds a year—and since I seemed to be aging at the same rate as everybody else, it added up. I managed to keep everything pretty much under control by using nonfat foods and cutting out some sugar—basically doing a low-fat diet. Nevertheless, by fall 1997, I had crept up to around 200 again.

To me, then, The Engineer's Diet was not as completely wacky as it might seem to others. Also, I recently remembered that Thomas Jefferson skipped lunch (according to the guide at Monticello). Jefferson lived into his mid-80s and was certainly in the skinny quartile of our American presidents—and probably one of the healthiest.

So one day I gave the plan a try and found it relatively easy to do. That first day, I felt a little bit of "routine" hunger at lunchtime; around noon or so, I had a cup of tea and the hunger pretty much went away. So I tried it for a few more days. Essentially I combined it with my regular low-fat, high vegetable and fruit diet. I felt like I began losing right away, but I didn't have a scale so I had no idea how much. But I could feel a difference around my waist. Within a few more days I could actually take up another notch on my belt. The first time you do that, you're not sure whether you're just holding your stomach in or you've actually lost weight.

But right away I noticed that I wasn't feeling sleepy in that period right after lunch when I usually take a nap—around 2 p.m. That was something I'd been doing for about 15 years. Now this was nice! And this has been consistent. Whenever I skip lunch, I have energy that persists right on through till after 4 o'clock. (I do experience some kind of fairly strong change or drop in energy at that time.)

Within the first 15 weeks I lost 13 pounds. During the rapid weight-loss phase, I also ate less dessert. Now I frequently have something sweet in the evening. I'm in what I consider a maintenance phase and am not really trying to lose weight. I notice that as soon as I have a series of lunches for a week or so, the weight starts to come on again. If I go right back to no lunches, it comes off.

Rebecca S.
Female, age 48

Progress after 13 weeks—weight loss 12 pounds

I've always been a starver. My way of losing weight was to stop eating. It was a lot easier than moderating or selecting foods that didn't add fat. And starving has some side benefits—like making you feel kind of high.

Also, starving works so fast—a couple of days on less than 500 calories a day and the pounds start falling off.

Last time I starved was in December 1995, right after a friend of mine died of AIDS. I pretended that I was doing the Weight Watchers "food-groups" plan, but actually, I just reduced calories to around 500–700 max. It felt great for awhile. I went to Europe in the spring and "binged" to comfort myself after that period of deprivation. By the time I came home, at least 8 pounds were back, and soon after, I weighed even more than before I starved. I stayed that way—and depressed about it—for more than a year.

I saw what Tom was doing and started skipping lunch toward the end of summer 1997. It took no time to get used to it (remember I do non-eating pretty well!) and within 13 weeks I lost 12 pounds. I did exactly what he suggested, which was to eat a light breakfast, skip lunch, have a juice before dinner, eat a nutritious, not particularly high-calorie dinner, and eat an apple before going to bed.

The greatest thing of all is the change in my energy level. It still blows my mind. I get up in the morning at a certain energy rate—which has always been high—and I just stay that way the whole day until after supper. It's incredibly helpful for someone who is running her own business. My whole life is about self-scheduling and self-discipline! Having this "no-brainer" eating schedule, and this steady energy, has been the miracle of time management I've been praying for. And I am not exaggerating one bit.

I've basically been on the plan ever since. At times, like when I was recently in Spain, I've had a few days of lunches. Guess what? I got tired much earlier on those days. And I noticed the scale stopped moving. At times, also, I've been giving in to social and boredom pressures and increased my sweets at night. When I do that, I can really see the difference—actually feel it—because the sugar makes me sluggish and makes me want more sugar. But I haven't gone even halfway back up to my original weight. And when I stop the sugar, I keep on losing. You couldn't pay me to go back to eating three times a day.

Barbara K.
Female, age 55

Progress after 16 weeks—weight loss 32 pounds

First of all, let me state that I am over age 55, and I have years of extra fatty tissue that is not enhancing my "good" looks. When I started this weight management program, I was basically inactive and lazy when it came to exercise and could always find an excuse not to. Most importantly, and to put it bluntly, I was not overweight, I was *fat*.

I wanted to lose weight, but I blamed everything on my hormones. After all, having gone through the big "change," I knew my body was not going to "change" anymore. The important point here is that I knew I had to lose weight, but how?

When Tom Shaw said that all it would take for me to get to my desired weight was to *not eat* lunch, I looked at him like he had two heads. He told me it would take time and that I wouldn't be able to see a drastic change right away. He said he thought I could lose a pound or 2 the first week and that within a month I could expect to lose 4 or more—every month from then on. On this schedule, it would take not even a year to lose the 50 pounds I wanted to get rid of.

I had noticed that Tom had lost a lot of weight and was keeping it off. He told me it was just from not eating lunch. Well, if he could do it, so could I! Still skeptical, I picked a Monday to start and ate breakfast. Let me stop here and say that for many years, I lived alone and seldom ate breakfast except on weekends. I always had a large lunch and seldom even ate dinner. What a change this was going to be for me. Thank goodness, it was early fall; the weather was warm and beautiful, and I was able to get out at lunch and drive by the lake.

That first day, I was very hungry soon after breakfast and already worried about dinner. It was only 10 a.m. and I was absolutely frantic. At this rate, I could see myself going home at night and eating everything in the house! I drank a cup of coffee and, yes, by lunch, I was verrrry hungry—but I did not eat. That afternoon, I drank a Pepsi (not diet), and by the time I got home, I had a terrible headache. I was hungry, crotchety, and prepared to eat anything in sight. But I had a piece of cheese with a few crackers and fixed dinner. By the time I sat down, I was still good and hungry, but not starving.

The second day was a little like the first; the third and fourth got a little bit better; but by the end of the week, I had, in fact, lost 1 pound. I knew I was on the way. I could go into detail about all the weeks since then, but it's enough to say that by not eating lunch every day, I have been able to lose 5 to 8 pounds every month. And the most surprising part is that it stays off! During the holidays, with all those good things to eat, I did not gain back anything. But I was able to splurge and eat things that would normally put a dieter out of control.

Sue A.
Female, age 45

Progress after 10 weeks—weight loss 10 pounds

I've been trying to lose 10 pounds for 10 years. My kids were born in 1987. Up until this year I had never been able to get back to my pre-pregnancy weight. My husband was always thinking about better ways to do things, so you can imagine how many strange experiments we tried. I was shopping for him and also buying altogether different food for the kids. I didn't think it was important to eat for myself, so I'd prepare his stuff, and eat some of it, and then cook for the kids, and eat some of that too. And during all this time, what I really wanted was my very own dinner!

I worked in an office job for several years. Every day I used to take a slice of turkey between two pieces of white bread smeared with mustard. I ate it like clockwork, even after it started tasting like paper towels. So, I always ate, either this meager sandwich from home, or restaurant fare, but never skipped lunch. It never occurred to me that you could go without eating. There was a mandate to eat just because it was noon. It didn't matter if I wasn't really hungry.

Things have really changed since I went on The Engineer's Diet. Physically, I have more energy. I don't get sleepy anymore during the day. The dilemma about what I'm supposed to do at noon is over. And so is the problem of preparing meals at home. It's important to me to eat with the kids, and now I can eat what they do. The fact is, I can eat whatever I want now—guilt free—and know that I'm not overdoing it. And my thinking has changed. Whenever I see crowded restaurants at noon, I find myself wondering why all those people with desk jobs are in there eating so much. I guess I think only those folks who really burn calories—by laying brick, or digging ditches, or mixing concrete—ought to be chowing down like that!

Also, I used to feel that I had to exercise to exhaustion. Now I know that I can exercise reasonably, and that I'm not dependent on exercise to lose weight. The Engineer's Diet is a reasonable diet and gives me a sensible exercise program. And by the way, I now keep my eye out for eclectic little corners and shops where the atmosphere and the coffee are great. It has given me a new kind of socializing that I really like.

Bob S.
Male, age 48
Progress after 8 weeks—weight loss 6 pounds

I was introduced to the "No Lunch" Diet by Tom Shaw in mid-October 1997 and followed this simple diet plan for about 8 weeks, losing about 6 pounds. This diet was so easy to follow, and the reasoning why it would work was so obviously correct, that I was successfully on the plan from the first day. I would work through lunch and whenever I thought of food, I would think about the fabulous dinner I was going to have. I would occasionally have a business lunch but as long as I didn't lunch regularly I knew that I would continue to lose weight. I returned to overeating—by eating lunch—with the Christmas holidays and gained over

10 pounds. I am now returning to The Engineer's Diet as a better way of eating.

My youngest son, Peter, also liked the idea of eliminating lunch to eliminate overeating. It made good sense to him, and he's 16 years old. He stopped eating lunch in school for about 4 weeks but his mother coerced him back into eating lunch like everybody else. His mother follows the old philosophy of overeating to stay healthy.

Victoria S.
Female, age 52

I don't like to weigh myself. I don't believe it really does any good. So when I quit eating lunch, I just quit and didn't really treat it like a diet. Besides, I think that when you're involved in a slow, steady change like that, you can get discouraged at the beginning if you have big expectations. The other thing is that some people tend to get anxious at the beginning and maybe overeat at the other two meals. Anyway, the best way for me was to not worry about it and not weigh myself all the time. So after the first two weeks, even though I couldn't prove it on the scale, I know I lost because my clothes started fitting a little better.

Eating this way has made me feel encouraged and more cheerful about losing weight. Eating doesn't preoccupy my mind all day like it used to. The pressure's off because I know I'm going to eat breakfast every morning and dinner every night. So I stop worrying. And there are no real surprises about mealtime. It gets pretty routine and that helps you relax about your weight. But I especially like it because, even though I have gotten off the diet a few times, I get right back on since I'm not discouraged. I also noticed that even when I didn't follow it, because I had already been doing it for a couple of weeks, my ideas changed about how much food I have to eat at any one meal. I can really tell when I'm eating something I don't need because my body has already adjusted to needing less food.

Alejandro C.
Male, age 27
Progress after 5 weeks—weight loss 5 pounds

This diet is great. My two favorite things about it are that it saves me money and I have more energy. I'm always careful about how often I go out to lunch anyway—to save money. Now I even save it on my grocery bill—a couple of bucks daily on average. And during the day, I don't have those energy slumps. Before, when I ate lunch, I always felt like going to sleep afterward.

I'm losing a pound a week, and I'm not even 100 percent regular about it. When I'm out of town and traveling, it's tougher than at home—kind of two steps forward and one backward. But it used to be the other way around. This is the first time I've ever lost weight without suffering. And I don't worry about eating lunch sometimes, because it's the exception now instead of the rule. If you keep in mind that this diet is a lifestyle change, then you can eat a lunch now and then and not have to worry.

I never used to eat breakfast, and then I'd pig out at lunch. Now I'm diligent about breakfast, and I really enjoy dinner. And it's not hard to go without food during the day; it's just that everybody else wants to go to lunch. But I can go out with a group and not eat; it's pretty easy. When people ask, I tell them about The Engineer's Diet. And they're interested. Because we're not talking about electrolytes and proteins and all that crap. This diet appeals to everyone's sensibilities because it makes sense. Besides I get to eat what I want at dinner, and I really enjoy it. People can see that when they see me eating dinner because I have a big smile on my face.

Manto Press

1442-A Walnut Street, Suite 390, Berkeley, CA 94709

Manto was a Theban woman
and an oracle at Delphi from
whose name comes the term
mantic speech, meaning
prophetic words.

Editorial
Rebecca Salome
Elizabeth Shaw
Sonya Manes

Design and Production
Harrison Shaffer

Illustration
Don Berry

PRINTED IN U.S.A.